Identity and the Natural Environment

Identity and the Natural Environment

The Psychological Significance of Nature

edited by Susan Clayton and Susan Opotow

The MIT Press
Cambridge, Massachusetts
London, England

This book was set in Sabon by SNP Best-set Typesetter Ltd., Hong Kong. Printed and bound in the United States of America.

Library of Congress Cataloging-in-Publication Data

Identity and the natural environment : the psychological significance of nature / edited by Susan Clayton and Susan Opotow.
    p. cm.
  Includes bibliographical references and index.
  ISBN 0-262-03311-9 (hc : alk. paper) – ISBN 0-262-53206-9 (pbk. : alk. paper)
  1. Environmental psychology. 2. Identity (Psychology). I. Clayton, Susan. II. Opotow, Susan.
  BF353.I34 2004
  155.9'1—dc21

                                                 2003051378

Printed on recycled paper.

10 9 8 7 6 5 4 3 2 1

Contents

Acknowledgments

This book has benefited from many conversations with colleagues at conferences and more informal venues. We would like to extend particular thanks to Mitch Thomashow, who helped to clarify our thinking; Clay Morgan and Sara Meirowitz at The MIT Press, who shepherded the book through the publication process; Faye Crosby, the late Edwin Giventer, Stu Oskamp, Dan Perlman, Carol Saunders, and Ethel Tobach, for their vital advice, support, and encouragement; Louise Chawla, for her comments on an earlier draft; and John Sell, for translations from the German. We also thank the contributing authors for what we learned from their chapters and for their good cheer. Finally, we have been inspired and energized by our children—Jay Clayton, Colin Howell, Nathan Chang, and Vera Chang—and we dedicate this book to them.

Identity and the Natural Environment

1

Introduction: Identity and the Natural Environment

Susan Clayton and Susan Opotow

Like many others, both of us value the environment, and this value has come to shape our professional and personal lives. For the past 15 years, we have worked as social psychologists to understand the deep-seated but unexamined beliefs and values people have regarding the natural environment (Clayton & Opotow, 1994; Opotow & Clayton, 1998). We share an abiding interest in the psychology of justice, and in our own research explore how people think about issues of fairness with regard to the natural environment (e.g., Clayton, 1996, 2000; Opotow, 1993, 1996).

Because we live far apart, we meet to exchange new ideas and enjoy the camaraderie of collaboration. One evening, over dinner at a vegetarian restaurant in Ann Arbor, Michigan, we discussed the powerful way that fairness and identity interact in environmental conflict. After dinner we saw our ideas crystallized with bracing directness in a graffiti exchange on the wall of the women's restroom. A first comment admonished: "Eat organic—no poison food. Love Earth—don't poison your home." A crude response jeered: "Eat shit you tree hugging faggot." This call and response captured a reality that we had observed both anecdotally and in our work on environmental perspectives in citizens' Wise Use environmental groups (Opotow & Clayton, 1998): People are impassioned about environmental issues, their environmental beliefs can affect other aspects of who they are, and environmental positions perceived as different than one's own can elicit a violent reaction.

The graffiti exchange prompted us to think about what drives such statements. They are not primarily about justice, although they have implications for what is seen as fair. Nor are they simply expressions of attitudes. Although the first writer was clearly taking an attitudinal stance, the second writer probably would not have responded as vehemently to a more factually based command. This exchange was more personal. The first statement evoked an emotional relationship between an individual and the natural environment. The second statement not only stereotyped the first writer, but also demonstrated a hostile, emotional, and aggressive reaction to that stereotype. We understood this exchange to be about identity, but identity in a larger and more powerful way than it is usually defined. This exchange placed our relationship to others within the larger context of how we view our relationship to the natural world.

Why does it matter if the natural environment and environmental issues have ties directly to our core identity? As the graffiti exchange illustrates, issues that are relevant to our sense of self attract attention, arouse emotion, and connect to other aspects of our life more than issues that are less personally significant (Kihlstrom et al., 1988). Concerns about self-image and self-presentation are also a strong motivation for behavior (see Baumeister, 1998; Onorato & Turner, 2001).

Understanding identity and its role in mediating behavior toward the natural world not only has provocative implications for research, but it also has important practical implications. If we better understand what makes people passionate about the environment, we can understand the psychological mechanisms capable of fostering protective environmental policies and behavior. At a local level, researchers in many countries have begun to find that environmentally sustainable behavior requires a strong community identity because both personal and collective identity determine whether the values of sustainability are adopted (see Pol, 2002; see also Van Vugt, 2001). At the global level, analyses of intransigent political conflicts suggest that they too are often fundamentally conflicts over identity and such crucial environmental resources as water (Matthews, 1999).

Identities describe social roles, and roles entail responsibilities. An environmental identity—how we orient ourselves to the natural world—can describe the way in which abstract global issues become immediate and personal for an individual. An environmental identity also prescribes a course of action that is compatible with individuals' sense of who they are.

The topic of identity and environment is only beginning to attract research from a wide variety of subdisciplines. This book brings some of this work together in order to capture its variety, energy, and relevance, and to impose coherence by placing the work within a descriptive framework. Our goal is to present empirical research and theory on the ways in which identity matters in determining human responses to the natural environment.

Background

To date, environmental scholarship has given insufficient consideration to the deep connections between identity and the natural environment. Empirical research on environmental topics frequently poses questions as objective matters for natural science: Is the average global temperature increasing? What are the likely effects of the destruction of rainforests on the extinction of species? Because environmental problems initially were seen as the result of technological advances that produced environmental degradation as by-products, environmental problems were also seen as requiring technological solutions.

There has been increasing recognition that environmental degradation is not purely a technological question, but is partly behavioral and attitudinal as well. This recognition has focused more attention on the ways in which people think about the environment (see, e.g., Geller, 1992; McKenzie-Mohr & Oskamp, 1995; Oskamp, 2000, 2002; Zelezny & Schultz, 2000.) Even social scientific research, however, often conceptualizes the human relationship with nature as a disinterested one. People are construed in economic rather than affective terms—as caring for the environment primarily because it furnishes us with resources. The majority of the research considers factors associated with sustainability, focusing on how people can be induced to make personal sacrifices for the environment through recycling or reducing their resource use. Along similar lines, economic research has estimated the value of the environment in terms of the financial sacrifices people will make for it (Knetsch, 1997). These approaches have obvious pragmatic as well as theoretical importance, but they provide an incomplete picture. Focusing on economic tradeoffs and concrete physical payoffs has significant limitations in attempts to change or understand behavior. As with the dilemma of the commons(Hardin, 1968), environmentally destructive behavior may be a short-term, rational choice for an individual or a small group, even when it has counterproductive outcomes for the long term and for the larger collective.

Research as well as intuition and anecdote indicate that the natural environment has value beyond its immediate or potential utility. Economic analyses are increasingly complicated by findings that people do not calculate the utility of nature in an economically logical way. Instead, people use environmental behavior and contributions to environmental organizations as a way to make statements about their personal and collective values—to define who they are through the causes they support (Ritov & Kahneman, 1997). Surveys have shown that environmental values have strong religious and/or moral overtones for many American citizens (Kempton, Boster, & Hartley, 1995); for example, a 2000 survey of the voting public found that 64 percent agreed that protecting the environment is a moral issue involving beliefs about what is morally right or wrong (Henry J. Kaiser Family Foundation, 2000). Cross-cultural studies find that environmental degradation is a source of concern for people across countries, social classes, and ethnicities (e.g., Kahn, 1999; Morrissey & Manning, 2000; Parker & McDonough, 1999; Rohrschneider, 1993).

Almost every day news items describe environmental conflicts that are about more than economic interests: questions of whether the right to enjoy unpolluted natural resources can override private property rights; whether the interests of nonhuman species count as much as the interests of humans; and why people persist in buying sport utility vehicles in the face of mounting evidence about global warming. A close look at these stories reveals people who are passionate about the issues, moved to tears at the loss of their ranch or to anger at the plight of endangered species, pleading for the protection of a way of life or demanding the restoration of a canyon. These emotions are not limited to people identified as "environmentalists"; those who are defined as being on the "antienvironmental" side of the fence—ranchers, hunters, and property owners—also argue their love of the land and their desire to preserve their relationship with it. As a farmer quoted in chapter 13 in this volume said, in defense of his property rights, "you can have my water when you peel my cold, dead hand off my pump."

If the value of nature is not purely economic, what is it? What makes our relationship with nature an emotional one? And what makes an environmental issue evoke strong moral considerations? We believe that the answer has to do with identity—how we define ourselves, others, and nature. This book is premised on the idea that people have strong feelings about nature because of its implications for both social and environmental identity. In chapter 10, a leader of an inner-city tree-planting program in Detroit puts it this way: "it just makes a difference, your environment, how you act and how you feel about yourself."

What Is Identity?

Identity is a concept with broad meaning, traditionally linked with self-concept and involving beliefs about who we are and who we want to be. However, there is little consensus about what identity is. Ongoing debates concern the extent to which identity is primarily single or multiple, independent or interdependent, personal or social (Ashmore & Jussim, 1997; Cross & Madson, 1997). In a psychoanalytic sense, identity is formed by separation: the developing child builds a sense of who it is by distinguishing itself from what it is not (Segal, 1973). Identity is also considered to be a product of social appraisal. We form a sense of ourselves based on the information we receive about ourselves from others (Cooley, 1902; Mead, 1934). Although identity, as a description of oneself, may be intuitively felt to be a stable personal attribute, even such enduring characteristics as gender and ethnicity are subject to situational and cultural variations that affect what is salient and how it is interpreted (Nagel, 1996).

Personal identity emerges in a social context that includes interpersonal and group memberships. This perspective emphasizes cultural aspects of identity and takes account of social interdependence (Snyder & Cantor, 1998). Identity in this context is not stable, but is layered, complex, and changing as it is negotiated in social interactions and conflicts (Carbaugh, 1996). As Martha Minow (1997) describes identity, it is more salient when it becomes fluid, such as when individuals or groups undergo geographical, social, and psychological shifts.

A conceptualization of identity in a changing social context is more complex than one that is static or purely intrapsychic, but it nevertheless remains a largely anthropocentric construct, rooted in multiple levels of social relationships. These analyses of identity miss the larger, non-human context within which all human relationships occur. In 1960

Harold Searles argued that a human relationship with the natural world is transcendentally important and ignored at peril to our psychological well-being (Searles, 1960). (See chapter 2 for a further discussion of Searles.) Yet psychological research has given scant attention to our relationship with the natural world. A broadened conception of identity would include how people see themselves in the context of nature, how people see animate and inanimate aspects of the natural world, how people relate to the natural world as a whole, and how people relate to each other in the context of larger environmental issues.

What Is Nature?
Like "identity," definitions of "nature" and "the natural environment" are complex and contested. The predominant meaning has traditionally been "our nonhuman surroundings" (Simmons, 1993, p. 11) with an understood dichotomy between what is a result of human influence and what remains untouched. The dichotomy between the natural and the manufactured is, of course, artificial. Nature has long been subject to human influence through what is planted, supported, or tolerated, and what is exterminated either directly or through elimination of its habitat. McKibben (1989) has famously made the case that there is no more "nature" in the traditional sense: "By changing the weather, we make every spot on earth man-made and artificial" (p. 58). Thus the idea that the experience of nature is separate from social experience is misleading. This myth may promote the idea that the preservation of the natural environment is irrelevant to the life of the average citizen, and contribute to the perception that environmentalists are misanthropic (see Morrison & Dunlap, 1986). In a book on Henry David Thoreau, David Botkin (2001) writes "This sense of isolation from nature reinforces the idea, prevalent today, that nature is 'out there' and that preserving nature is generally an activity that takes place over the horizon. . . . As a result, the conservation and protection of wild living resources is typically seen as an activity that is beyond our day-to-day urban experience" (p. 94). In this book, we use the terms *nature* and *the natural environment* in the average person's sense, to refer to environments in which the influence of humans is minimal or nonobvious, to living components of that environment (such as trees and animals), and to nonanimate natural environmental features, such as the ocean shore. Weigert (1997, p. 49) has

referred to this as the "relatively natural environment." However, we emphasize that the experience of nature can take place in urban settings as well as in remote wilderness areas.

Identity and Nature

Scholars from a variety of disciplines have begun to explore the relevance of identity to the natural environment. Philosopher Carolyn Merchant (1992) has described different ways of valuing the environment that include extending the boundaries of what is important to oneself alone (egocentric), to humans in general (anthropocentric), and to the biosphere (ecocentric or biocentric). Psychological research (e.g., Axelrod, 1994; Stern & Dietz, 1994) has affirmed the significance of these different value orientations in predicting environmentally sustaining behavior. Biologist E. O. Wilson (e.g., 1984) has proposed a genetically based tendency to affiliate with nature, "biophilia," suggesting that the relationship with the natural world is a hard-wired part of human nature. Stephen Kellert (e.g., 1993, 1997) has elaborated the ways in which the biophilia tendency might be expressed, and has detailed the adaptive benefits that accrue even today from what he describes as the "link between personal identity and nature" (1993, p. 43).

Some have moved from theory to advocacy in arguing that people need to connect with nature at a deep and personal level, even to redefine themselves in a way that includes the natural world. A philosophical perspective known as "deep ecology" (Naess, 1989; see also chapter 2 in this volume) describes this as necessary in order to live a life that is in balance with nature. "Ecopsychology," similarly, represents a therapeutic orientation which holds that humans need to rediscover their ties to the natural world in order to experience full mental health (Roszak, 1992; Thomashow, 1998; Winter, 1996).

Our own work exploring distinct conceptions of justice with regard to environmental issues (Clayton, 2000; Clayton & Opotow, 1994) led us to the question of identity because beliefs about what is fair are fundamentally entwined with who we are, how we relate to others, and what that means in terms of rights and obligations (Clayton & Opotow, 2003). Opotow (1987, 1990) has argued that we distinguish persons and environmental entities with moral worth from those without such worth by excluding the latter from considerations of fairness, justice, or the moral community. In other words, identity underlies beliefs about deserving and fairness. Research by Clayton and others (e.g., Clayton, 1996; Kahn, 1999) strongly suggests that nature and natural entities (trees, species, ecosystems) are given moral consideration by individuals who care about environmental issues. (See also Nash, 1989, and Stone, 1974, for historical and legal perspectives on the moral standing of nature.) Several writers have introduced terms that are similar to the focus of this book. Thomashow (1995), for example, has written about "ecological identity," meaning, in part, "the ways people construe themselves in relationship to the earth" (p. 3). Similarly, Weigert has written about "environmental identity," meaning the "experienced social understandings of who we are in relation to, and how we interact with, the natural environment as other" (1997, p. 159). Others have described an "environmental self" or an "ecological self" (e.g., Cantrill & Senecah, 2001; Naess, 1989), but none of these terms has succeeded in claiming definitional turf or achieving consensus about its meaning (see Neisser, 1997 on the ecological self). This lack of consensus on terminology reflects the slipperiness of the concepts. When even "nature" and "identity" are hard to pin down, it is difficult—and not necessarily desirable—to construct a rigid definition of environmental identity. The fact that the concept has been approached from a number of disciplinary and theoretical perspectives, however, reflects growing interest in the intersection between identity and environment as well as awareness of its practical significance.

Environmental Identity Elaborated

We propose an integrative construct of environmental identity that encompasses multiple meanings as well as a recognition of the dynamic nature of identity. One way of thinking about environmental identity concerns the way in which we define the environment, the degree of similarity we perceive between ourselves and other components of the natural world, and whether we consider nature and nonhuman natural entities to have standing as valued components of our social and moral community (Opotow, 1993, 1996). For example, pretechnological cultures sometimes ascribe identity to natural forces and objects such as trees, animals, mountains, or winds, endowing them with intentionality, emotional response, subjective perspective, or simply spiritual significance (see chapter 9). Although such conceptions of nature are considered unscientific anthropomorphism by current Western standards, they may reflect a more sophisticated understanding (see, e.g., Nelson, 1993) based on a perceived similarity rather than equivalence (cf. Kahn, chapter 6 in this volume, and Kalof, chapter 8).

Not only is the natural world given an identity through the way in which people view and experience their relationship with it, but it also influences individual identities. Merely by existing as an important symbolic, physical, and political reference point that is encountered in books, stories, public debates, and experiences, the natural environment serves to inform people about who they are. Consistent with psychological research on "place attachment" (see Altman & Low, 1992), emotional connections to particular environmental aspects of places people have lived—rocky terrain, harsh winters, and the ocean shore—serve to shape individuals' self-definitions. Similarly, people may define themselves through the ways that they interact with nature as environmentalists, hikers, and/or landowners; or through the relationships they form with animals (see Plous, 1993) and their perception of being both similar to and different from animals (see chapter 8).

These strong attachments to and contrasts with nature can contribute to the formation of group identities in environmental contexts. Bird watchers, urban dog walkers, hikers, and hunters develop bonds of friendship as well as engaging in collective action, battling private development and public initiatives as they advocate land protection or fight hunting restrictions or leash laws. Thus nature-oriented activities can elicit strong social connections that take on intensified meaning in environmental conflicts between those who want to interact with nature in one way and those wishing to interact in another. Those working to preserve a forest, for example, become "us," while those who log the forests become "them." As the graffiti exchange illustrates, stereotypes that are attached to particular environmental values, attitudes, and behaviors can play out as social and political conflict.

A Model of Environmental Identity
We propose that environmental identity can be usefully conceptualized as occurring along a dimension anchored by minimal and strong levels

of social influence. Environmental identities inevitably contain a social component because they depend on and ultimately contribute to social meaning. How we understand ourselves in nature is infused with shared, culturally influenced understandings of what nature is—what is to be revered, reviled, or utilized. Social variables affect how much we are able and choose to focus on the natural environment and how we interpret what we see. However, although social influence is inevitable, the degree of influence varies.

With an environmental identity that is minimally influenced by social factors, individuals or groups view themselves as experiencing and understanding nature directly, with little or no social mediation. Nature is seen as distinct and apart from social living, and social conflicts and group memberships are less prominent (table 1.1). In contrast, with an environmental identity that is strongly influenced by social factors, individuals or groups view themselves as situated within potent social categories, in which political realities, activism, and social conflicts are prominent. Described in terms of the gestalt figure–ground shift (Kohler, 1959), at times the natural environment is the figure (the main focus of a picture that is central in importance) while the social context is the ground (peripheral elements that serve as background). A shift in perspective, however, can make social groupings into the figure and the environment merely background (figure 1.1). Between these two perspectives are conceptions of environmental identity in which the social and natural environment exert a roughly equivalent influence; they are interdependent.

Table 1.1
Environmental identity

Influence of social identity	
High	Low
Social group identity is prominent	Individual experience is prominent
Positions self in a social world	Positions self as interacting with nature
Political reality is influential	Nature is seen as apart from human activity
Group focus	Individual focus
Social conflict prominent	Social conflict less prominent

Figure 1.1
Social and natural influences as figure and ground in environmental identity.

Although this depiction of environmental identity is static, in actuality, environmental identity is complicated by a dynamic interplay between the social and environmental. As this book's chapters illustrate, figure and ground continually encroach on and envelop each other as change is generated by growth, understanding, experience, and conflict. For example, as a consequence of environmental experiences such as tree planting or exposure to wild animals in a zoo, an environmental identity that was initially rooted in social, community experiences can expand to include the natural world. Like Max's bedroom in the story *Where the Wild Things Are* (Sendak, 1963), which gradually morphed into a jungle, the environment can transform social identity, increasingly playing a larger role in one's sense of self and one's world view as a person in nature. Similarly, an environmental identity that had been attuned less to social forces and more to the rhythm, challenges, or beauty of the natural world, can become more politically and socially charged as threats to the environment are discerned. As social concerns intrude on what had previously been a more direct relationship with nature, environmental identity comes to include social issues, affiliations, and oppositions.

In sum, one's social orientation leads to ways to position oneself environmentally, while one's environmental orientation leads to ways to position oneself socially. Environmental identity, therefore, involves a dynamic interplay between what is perceived as central and as peripheral, with the social and environmental encroaching on and redefining each other.

Purpose and Structure of the Book

This book demonstrates that identity is relevant to understanding human interactions with the natural environment. All the chapters approach the topic of identity and the natural world from the perspective of social and behavioral science. In addition to offering theoretical advances for identity and environmental studies, the chapters address the practical and puzzling question of why concern about, and regard for, the natural environment does not always lead to action to protect it. The diverse methodologies, environmental contexts, and orientations toward human–nature interaction presented here offer perspectives that have relevance for policy makers, those involved in local and global environmental conflicts, developmental theorists, environmental educators, and people studying justice and values. By bringing together a variety of conceptual, contextual, and methodological approaches, we hope to suggest an overall structure for an emerging field and to propose fruitful directions for future research.

When we began this book, we sought agreement on a working term and asked the contributing authors about their preference for "environmental" or "ecological" identity. (See Thomashow, 1995, and Weigert, 1997, for earlier uses of these terms.) Instead of a consensus, we had a lively e-mail exchange in which the authors disagreed about the connotations of each term. Some authors preferred the term *ecological identity*, arguing that it better describes a sense of the self in relation to nature, or the self as part of an ecosystem, as well as avoiding the confusion caused by the fact that "environment" can include the built and even the social environment. Others preferred *environmental identity*, feeling that it has more intuitive meaning for the average individual, relating more clearly to what are known as "environmental issues." While we (Clayton and Opotow) prefer environmental identity and use it in this introduction, it is characteristic of the field that little consensus exists. Our intention is not to assign a single, fixed definition to environmental identity, but to bring together a range of conceptualizations in order to better understand the many ways that environmental and social identities entwine.

Research Methods

The study of identity and the natural environment presents logistic and methodological challenges that require diverse and creative approaches. It is important not only to understand a particular environmental context and the ways people respond in that context but also to attain internal and external validity in order to develop generalizable conclusions. The chapters in this book meet these criteria by utilizing multiple research approaches. Qualitative methods produce rich data with powerful expressions of individuals' responses to environmental issues. These approaches include structured and semistructured interviews, focus groups, case studies, participant observation, and analysis of archival material. Other studies provide more experimental control by utilizing such quantitative methods as questionnaires and experimental research.

The chapter authors have examined a wide range of environmental contexts and groups. The latter cover a broad range of ages, socioeconomic backgrounds, and countries of residence. The studies consider environmental identity for children and adults, rural and urban dwellers, ranchers and biologists. They also consider environmental protection and degradation at several levels of analysis—from the loss of a particular field or tree at one end to global climate change at the other. Just as environmental changes can be understood at different levels, identity has to be understood at these levels as well. As the authors ably show, nature is significant for individuals, communities, and regions. Taken as a whole, this book depicts a variety of ways in which individual and societal constructions of identity influence the ways we think about, relate to, and behave toward the natural world.

Organization of the Book

Steven J. Holmes begins the consideration of environmental identity by reviewing the extant literature relating to identity and the natural environment and by discussing the intersection of nature and identity in earlier theories and in the lives of other writers. Reflecting on the field as a whole, Holmes raises new questions about the complex relationship between social and environmental identities in the context of environmental awareness, conflict, and change.

The remaining chapters all report original empirical research. They are divided into three parts based on the degree of social influence on environmental identity: minimal, moderate, or strong. In the work discussed in the first part, a direct relationship to the natural world is more influential than social aspects of environmental identity. In the second part, direct and social aspects of environmental identity are roughly in balance. In the third part, the social aspects of environmental identity are more influential than a direct relationship with the natural world.

Part I. Experiencing Nature as Individuals   These chapters describe environmental identity in abstract and holistic terms. Social forces, although relevant, are ground rather than figure; the focus is on the way that environmental identity emerges through immediate and personal experiences with the natural world that change people's understanding of themselves and nature.

The authors consider the way in which individuals construct the "identity" of nature, as well as ways in which environmental experiences foster awareness and an environmental identity in children and adults. A naturalist quoted in chapter 4 expressed this abstract, holistic, yet direct relationship with nature: "There's a feeling like you don't end at the tips of your fingers or the top of your head. Because what's going on has to do with what you're doing, and with other things that are going on, so you're kind of a part of this whole thing, and very small because there's so much more out there." This direct relationship with nature also seems to evoke ethical standards for human behavior toward nature based on a recognition of the macrolevel significance and standing of nature. As a child quoted in chapter 5 says, "trees have a right to stand there, and bushes, they have the right to grow." It then follows, as another child quoted in chapter 6 says, that "when you're dealing with what nature made, you need not destroy it."

Susan Clayton elaborates on the concept of environmental identity as a meaningful source of self-definition and discusses why the natural environment might be particularly relevant to the self. She presents research on the construction and validation of an environmental identity scale to examine individual differences in the salience or strength of an environmental component in the self-concept. Her research shows that environmental identity relates in a meaningful way to values, attitudes, and behaviors, and is predictive of the position people take with regard to environmental conflicts.

Gene Myers and Ann Russell discuss the ways in which an environmental identity can emerge through interaction with the natural, non-human environment, parallel to the ways in which identity emerges from social interaction. They specify criteria for claiming that interactions have relevance for identity, and draw on in-depth interviews with ten men who have interacted with black bears to illustrate the way in which these criteria are met. Myers and Russell point out that even identity formed through interactions with animals is partly mediated through socially determined interpretations.

Ulrich Gebhard, Patricia Nevers, and Elfriede Billmann-Mahecha utilize focus group discussions to explore environmental reasoning in detail, and find that children use human identity to construct an anthropomorphic identity for such natural objects as trees. The authors contend that this anthropomorphic reasoning allows the children to feel empathy with nature and endow it with moral standing. A greater use of anthropomorphic reasoning among younger, compared with older children, suggests the influence of socialization on the way in which we think about nature. The authors discuss the relevance of their research for designing environmental education programs.

Peter H. Kahn, Jr., describes cross-cultural research on environmental moral reasoning. His findings suggest that the natural environment is important to children and that they consider it to have moral standing. Kahn finds two principal ways in which children think about environmental issues: anthropocentric, in which the environment is merely a source of human value, and biocentric, in which nature is given moral standing. The development of biocentric reasoning suggests that there may be a maturational component in the way we think about our relationship with nature. Kahn suggests that the disconnect between attitudes about nature and environmental behavior is a consequence of the fact that people have multiple identities that vary in significance from one situation to another.

Elisabeth Kals and Heidi Ittner report research on children's motives for nature-protective behavior. According to Kals and Ittner, environmental identity in children is evinced by a combination of emotional attachment to nature and moral concern about threats to nature. Their studies of German and British children show that even young children have developed a view of the natural environment based in part on emotional reactions and moral reasoning, and that this view affects their attitudes toward environmental protection.

Part II. Experiencing Nature in Social and Community Contexts   In these chapters social and natural forces exert an interdependent effect on environmental identities. Experiences and understandings of nature are fundamental, but one's relationship to nature is understood as pervasive and ordinary, resulting from the specific sociocultural practices of everyday life. The studies elaborate some of the intervening levels of analysis that structure and moderate the human–nature relationship. An activist quoted in chapter 9 describes the relationship between social and environmental forces in the context of a tree-planting program: "We don't know if we're organizing communities to plant trees or planting trees to organize communities."

The authors examine the relationships between people and nature in such diverse contexts as people's perceived similarities to animals, people's connection to plants in their communities, and people's perceptions of nature in a coastal community. Ethical concerns about fairness and justice at this level are likely to be constructed at the microlevel, attuned to the fairness of procedures or distributions of resources within a specific neighborhood or locale, or specific instances of animal treatment.

Linda Kalof explores the relationship between how people think about themselves and how they think about animals and discusses the role of culture in constructing attitudes toward natural entities. Based on questionnaires, she finds that people think about themselves and animals in similar ways. Her research also examines more traditional, demographic features of social identity such as race and gender, finding effects of ethnicity but not gender on how animals are regarded.

Robert Sommer discusses how and why trees are significant to people. He reviews five theoretical approaches relevant to tree–human relationships and summarizes research on how involvement in tree-planting programs, compared with passive exposure to city trees, affects city dwellers' sense of personal empowerment and community identity. Sommer discusses the implications of the findings on tree–human relationships for future research and for the development of effective urban environmental programs.

Maureen E. Austin and Rachel Kaplan focus on the people who fill instrumental roles in implementing tree-planting programs: the leaders who initiate them, and the maintenance workers who ensure their success. Through interviews, the authors explore the ways in which individual differences in experience and knowledge affect involvement in the programs, and how involvement in such programs can transform the ways in which the participants think about themselves and their communities.

Volker Linneweber, Gerhard Hartmuth, and Immo Fritsche explicitly relate environmental identity to social position. In their study of the inhabitants of a threatened coastal zone, the authors found that perceptions of local environmental issues varied according to the social position of the respondent. They suggest that an environmental identity constitutes a set of cognitions about the environment that is relevant for a person in his or her specific social and geographic context. As such, environmental identity mediates the importance of environmental issues for individuals.

Part III. Experiencing Nature as Members of Social Groups   In the final part, environmental identity is defined within a strongly social context in which group affiliations and conflict play an influential role. Identity, whether chosen or ascribed, is partly determined by how individuals and groups position themselves with or against others on environmental issues. The perspective on environmental identity in these chapters is sensitive to the group, institutional, regional, and national levels, as well as the nuances of ecology that take account of multiple natural and human forces on nature.

The first two chapters describe groups that find themselves in opposition over specific environmental issues, while the final two chapters focus on people who self-identify as environmentalists, and who thus come into conflict with some of the values of the dominant social paradigm (e.g., consumption as a goal, economic considerations as primary). Issues of justice and fairness, which become overt in situations of conflict

(Lerner, 1981), are important for the individuals described in these chapters; procedure, outcome, and inclusion are particularly significant in regard to groups rather than individuals. Environmental issues are seen at least partly in terms of how social groups are treated, as illustrated by a rancher who asserts in chapter 12: "Don't ask me to be the only person that pays the burden. . . . If the population in general wants to protect endangered species, which I think is great, then the population in general should be willing to pay for it."

Susan Opotow and Amara Brook describe environmental identity in the context of rangeland conflicts. Their chapter connects to those in parts I and II by showing how social position can affect the moral standing given to a natural entity. On the basis of interviews with ranchers and surveys of landowners, they report on reactions to the endangered species listing of the Preble's meadow jumping mouse. Opotow and Brook show how both social and environmental identities can polarize ranchers and nonranchers, and how this can lead to a moral exclusion that delegitimizes the needs and concerns of those perceived as being on the opposing side. They discuss the implications for more constructive ways of handling environmental conflicts that take identity into account.

Charles D. Samuelson, Tarla Rai Peterson, and Linda Putnam examine the way in which group identities frame competing views of a dispute over water resources. They seek to understand the conditions under which knowledge motivates action in environmental issues, and propose that the answer is identity. They describe public participation in a watershed restoration project that attempts to overcome intergroup tensions. The authors argue that by facilitating communication under the right circumstances, this approach can encourage the development of a superordinate identity based on concern for the watershed and can increase the possibility of consensual action.

Stephen Zavestoski bases his chapter on the recognition that a committed ecological identity, although vitally important to at least a subset of the population, is not supported by society at large. Drawing on interviews with and observations of members of the deep ecology movement, he discusses the ways in which ecological identities have to be actively maintained through both social interaction and interaction with the natural environment. He defines ecological identity as a set of cognitions that could allow us to anticipate the environmental consequences of our behavior. Thus, when it is activated, an ecological identity necessarily motivates proenvironmental behavior.

Willett Kempton and Dorothy C. Holland draw on interviews with members of diverse environmental organizations to discuss how different types of environmental identity emerge. They explore ways in which self-identification as an environmentalist develops out of action. For Kempton and Holland, an environmental identity is composed of the importance of environmental issues, the identification of oneself as an actor, and knowledge. Identity and action then build on each other in a positive feedback loop.

Summary

Environmental conflicts will be neither understood nor constructively resolved unless we recognize the ways in which they reflect individual and group identities. Similarly, attempts to change behavior in a proenvironmental direction that ignore people's underlying environmental and social identities may have only a short-term effect; their behavior may revert to earlier forms when incentives to change are removed. Because environmental problems are increasingly important, and because environmental issues appear to engage moral reasoning and beliefs in a unique and powerful way, we need a better understanding of the connections between environmental issues and identity.

Each aspect of environmental identity—at minimal, moderate, and strong levels of social influence—can be understood as a different lens for looking at the human–nature relationship. Each person has the capacity to relate to nature at an abstract and a holistic level (minimal social influence), in practical and culturally grounded ways (moderate social influence), and in ways that are attuned to the complexity of multiparty environmental conflicts (strong social influence).

We can use these different lenses to examine why environmental identity does not always predict environmental action and understand when it is more likely to do so. When social influences are minimal, individuals who otherwise have a strong environmental identity may overlook specific societal threats to the ecosystem, as well as societal actions that can be taken to protect it. It is easy for individuals to talk about the rights of nature, as in some of these chapters, because there may be little sense of corresponding responsibilities. When social influences are moderate and a human–nature interaction exists within a specific social context, it can be easy to miss the big picture—a sense of how one's own, local environmental experiences can relate to global phenomena. Finally, when social influences are strong, political and social conflicts may lead to a focus on specific political details and strategies that overwhelm a more general concern for nature.

An awareness of environmental identity at each of these levels can reveal the different ways that people understand and respond to the psychological and moral significance of nature, as well as the practical ways in which that response can be translated into specific actions within a social context. Without minimizing the differences among the chapters in this volume, we suggest that they imply that proenvironmental action can be facilitated at three levels:

1. First, proenvironmental action will be facilitated when individuals see nature as an entity with moral standing rather than merely a source of resources to exploit. This should lead to a recognition that we have some responsibility to protect nature.

2. Second, proenvironmental action will be facilitated when social environments (both physical and conceptual) are designed to nurture a feeling of connectedness to nature and an awareness of the local impact of global environmental issues.

3. Third, proenvironmental action will be facilitated when social contexts support proenvironmental identities and encourage a recognition of shared concern for the environment that crosses and blurs existing group boundaries.

References

Altman, I., & Low, S. M. (Eds.) (1992). *Place attachment.* New York: Plenum.

Ashmore, R. D., & Jussim, L. (1997). Introduction: Toward a second century of the scientific analysis of self and identity. In R. D. Ashmore & L. Jussim (Eds.), *Self and identity: Fundamental issues* (pp. 3–19). New York: Oxford.

Axelrod, L. J. (1994). Balancing personal needs with environmental preservation: Identifying the values that guide decisions in ecological dilemmas. *Journal of Social Issues, 50*(3), 85–104.

Baumeister, R. F. (1998). The self. In D. T. Gilbert, S. T. Fiske, & G. Lindzey (Eds.), *Handbook of social psychology* (4th ed., pp. 680–740). New York: McGraw-Hill.

Botkin, D. B. (2001). *No man's garden: Thoreau and a new vision for civilization and nature.* Washington, D.C.: Island Press.

Cantrill, J. G., & Senecah, S. L. (2001). Using the "sense of self-in-place" construct in the context of environmental policy-making and landscape planning. *Environmental Science and Policy, 4,* 185–203.

Carbaugh, D. (1996). *Situating selves.* Albany: State University of New York Press.

Clayton, S. D. (1996). What is fair in the environmental debate? In L. Montada & M. J. Lerner (Eds.), *Current societal concerns about justice* (pp. 195–211). New York: Plenum.

Clayton, S. D. (2000). Models of justice in the environmental debate. *Journal of Social Issues, 56*(3), 459–474.

Clayton, S. D., & Opotow, S. V. (Eds.) (1994). Green justice: Conceptions of fairness and the natural world. [Special issue]. *Journal of Social Issues, 50*(3).

Clayton, S. D., & Opotow, S. V. (2003). Justice and identity: Changing perspectives on what is fair. *Personality and Social Psychology Review, 7.*

Cooley, C. H. (1902). *Human nature and the social order.* New York: Scribner.

Cross, S. E., & Madson, L. (1997). Models of the self: Self-construals and gender. *Psychological Bulletin, 122,* 5–37.

Geller, E. S. (1992). Solving environmental problems: A behavior change perspective. In S. Staub & P. Green (Eds.), *Psychology and social responsibility: Facing global challenges* (pp. 248–268). New York: New York University Press.

Hardin, G. (1968). The tragedy of the commons. *Science, 162,* 1243–1248.

Henry J. Kaiser Family Foundation (2000). *Washington Post* (Sept. 7–17). *2000 Election Values Survey.* Retrieved March 25, 2003 from Lexis Nexis.

Kahn, P. H., Jr. (1999). *The human relationship with nature: Development and culture.* Cambridge, Mass.: MIT Press.

Kellert, S. (1993). The biological basis for human values of nature. In S. Kellert & E. O. Wilson (Eds.), *The biophilia hypothesis* (pp. 42–69). Washington D.C.: Island Press.

Kellert, S. (1997). *Kinship to mastery: Biophilia in human evolution and development.* Washington D.C.: Island Press.

Kempton, W., Boster, J. S., & Hartley, J. A. (1995). *Environmental values in American culture.* Cambridge, Mass.: MIT Press.

Kihlstrom, J. F., Cantor, N., Albright, J. S., Chew, B. R., Klein, S. B., & Neidenthal, P. M. (1988). Information processing and the study of the self. In L. Berkowitz (Ed.), *Advances in experimental social psychology* (Vol. 21, pp. 145–180). San Diego: Academic Press.

Knetsch, J. L. (1997). Reference states, fairness, and choice of measure to value environmental changes. In M. H. Bazerman, D. M. Messick, A. E. Tenbrunsel, & K. A. Wade-Benzoni (Eds.), *Environment, ethics, and behavior* (pp. 13–32). San Francisco: New Lexington Press.

Kohler, W. (1959). *Gestalt psychology.* New York: Mentor Books.

Lerner, S. C. (1981). Adapting to scarcity and change (I): Stating the problem. In M. J. Lerner and S. C. Lerner (Eds.), *The justice motive in social behavior: Adapting to times of scarcity and change* (pp. 3–10). New York: Plenum.

Matthews, M. (1999). Politics of water stalls peace process. *Baltimore Sun*, Dec. 8, p. 1A.

McKenzie-Mohr, D., & Oskamp, S. (Eds.) (1995). Psychology and the promotion of a sustainable future [Special issue]. *Journal of Social Issues, 51*(4).

McKibben, W. (1989). *The end of nature.* New York: Random House.

Mead, G. H. (1934). *Mind, self, and society.* Chicago: University of Chicago Press.

Merchant, C. (1992). *Radical ecology: The search for a livable world.* New York: Routledge, Chapman, and Hall.

Minow, M. (1997). *Not only for myself: Identity, politics, and the law.* New York: New Press.

Morrison, D. E., & Dunlap, R. E. (1986). Environmentalism and elitism: A conceptual and empirical analysis. *Environmental Management, 10,* 581–589.

Morrissey, J., & Manning, R. (2000). Race, residence, and environmental concern. *Human Ecology Review, 7,* 12–24.

Naess, A. (1989). *Ecology, community, and lifestyle* (D. Rothenberg, trans.). New York: Cambridge University Press.

Nash, R. (1989). *The rights of nature: A history of environmental ethics.* Madison: University of Wisconsin Press.

Nagel, J. (1996). *American Indian ethnic renewal: Red power and the resurgence of identity and culture.* New York: Oxford University Press.

Neisser, U. (1997). The roots of self-knowledge: Perceiving self, it, and thou. In J. G. Snodgrass & R. L. Thompson (Eds.), *The self across psychology* (pp. 18–33). New York: New York Academy of Sciences.

Nelson, R. (1993). Searching for the lost arrow: Physical and spiritual ecology in the hunter's world. In S. R. Kellert & E. O. Wilson (Eds.), *The biophilia hypothesis* (pp. 201–228). Washington, D.C.: Island Press.

Onorato, R. S., & Turner, J. C. (2001). The "I," the "Me," and the "Us": The psychological group and self-concept maintenance and change. In C. Sedikides & M. Brewer (Eds.), *Individual self, relational self, collective self* (pp. 147–170). Philadelphia: Psychology Press.

Opotow, S. (1987). Limits of fairness: An experimental examination of antecedents of the scope of justice. Doctoral dissertation, Columbia University. *Dissertation Abstracts International*, DAI-B 48/08, p. 2500, Feb. 1988.

Opotow, S. (1990). Moral exclusion and injustice: An introduction. *Journal of Social Issues, 46*(1), 1–20.

Opotow, S. (1993). Animals and the scope of justice. *Journal of Social Issues, 49*(1), 71–85.

Opotow, S. (1996). Is justice finite? The case of environmental inclusion. In L. Montada & M. Lerner (Eds.), *Social justice in human relations: Current societal concerns about justice* (Vol. 3, pp. 213–230). New York: Plenum.

Opotow, S., & Clayton, S. (1998). What is justice for citizens' environmental groups? Poster presented at the meeting of the International Association for Applied Psychology.

Oskamp, S. (2000). A sustainable future for humanity? How can psychology help? *American Psychologist, 55,* 496–508.

Oskamp, S. (2002). Environmentally responsible behavior: Teaching and promoting it effectively. *Analysis of Social Issues and Policy, 2,* 173–182. http://www.asap-spssi.org/pdf/asap034.pdf.

Parker, J. D., & McDonough, M. H. (1999). Environmentalism of African Americans: An analysis of the subculture and barriers theories. *Environment and Behavior, 31,* 155–177.

Plous, S. (Ed.) (1993). The role of animals in human society [Special issue]. *Journal of Social Issues, 49*(1).

Pol, E. (2002). The theoretical background of the city-identity-sustainability network. *Environment and Behavior, 34,* 8–25.

Ritov, I., & Kahneman, D. (1997). How people value the environment. In M. H. Bazerman, D. M. Messick, A. E. Tenbrunsel, & K. A. Wade-Benzoni (Eds.), *Environment, ethics, and behavior* (pp. 33–51). San Francisco: New Lexington Press.

Rohrschneider, R. (1993). Environmental belief systems in Western Europe: A hierarchical model of constraint. *Comparative Political Studies, 26,* 3–29.

Roszak, T. (1992). *The voice of the earth.* New York: Simon and Schuster.

Searles, H. (1960). *The nonhuman environment.* New York: International Universities Press.

Segal, H. (1973). *Introduction to the work of Melanie Klein.* New York: Basic Books.

Sendak, M. (1963). *Where the wild things are.* New York: Harper & Row.

Simmons, I. G. (1993). *Interpreting nature: Cultural constructions of the environment.* New York: Routledge.

Snyder, M., & Cantor, N. (1998). Understanding personality and social behavior: A functionalist strategy. In D. T. Gilbert, S. T. Fiske, & G. Lindzey (Eds.), *The handbook of social psychology* (4th ed., pp. 635–679). Boston: McGraw-Hill.

Stern, P., & Dietz, T. (1994). The value basis of environmental concern. *Journal of Social Issues, 50*(3), 65–84.

Stone, C. D. (1974). *Should trees have standing? Toward legal rights for natural objects.* Los Altos, Calif: Kaufmann.

Thomashow, M. (1995). *Ecological identity: Becoming a reflective environmentalist.* Cambridge, Mass.: MIT Press.

Thomashow, M. (1998). The ecopsychology of global environmental change. *Humanistic Psychologist, 26,* 275–300.

Van Vugt, M. (2001). Community identification moderating the impact of financial incentives in a natural social dilemma: Water conservation. *Personality and Social Psychology Bulletin, 27,* 1440–1449.

Weigert, A. J. (1997). *Self, interaction, and natural environment.* Albany: State University of New York Press.

Wilson, E. O. (1984). *Biophilia: The human bond with other species.* Cambridge, Mass.: Harvard University Press.

Winter, D. (1996). *Ecological psychology: Healing the split between planet and self.* New York: Harper Collins.

Zelezny, L. C., & Schultz, P. W. (Eds.) (2000). Promoting environmentalism [Special issue]. *Journal of Social Issues, 56*(3).

2

Some Lives and Some Theories

Steven J. Holmes

The chapters in this volume take their place in a broad context, one that extends beyond environmental professionals to a variety of fields in the humanities and social sciences, and to the lived experiences of people. Indeed, significant experiences in and of the natural world appear regularly in the life stories of individuals from all walks of life, in informal personal reminiscences and storytelling, as well as in more formal literary and biographical works. Here I review some of the approaches through which the intersections of identity and the natural world have been explored by scholars and by writers, emphasizing the implicit dialogue between scholarly study and actual lives that is one of the hallmarks of this field. Without claiming to offer a comprehensive survey, I discuss relevant scholarship and life writing under three headings:

• *Developmental and psychoanalytic perspectives*, including the early work of Edith Cobb and Harold Searles, as well as more recent scholarship;
• *Place theory*, as formulated in humanistic geography and environmental psychology; and
• *The links among identity, ethics, and action*, including reflections by deep ecologists, phenomenological philosophers, and ecopsychologists.

I conclude by noting briefly what seem to me to be some of the most important continuing questions and challenges of the field, some of which are addressed by the following chapters in this book.

Developmental and Psychoanalytic Perspectives

Katy Payne, a wildlife biologist and writer, grew up surrounded by the broad fields and steep gorges of the Finger Lakes region of central New

York. Raised on a family farm and taught by *The Jungle Books*, *The Wind in the Willows*, and her father's self-created Johnny Possum stories that she was "of one blood" with the animals, her most powerful early identity was forged in close relationship with the natural world:

> I remember my first encounter with myself, on a high day in later summer. Standing alone in a field where wildness crowded up yellow and green against our garden and house, I said out loud, "This is the happiest day of my life and I'm eleven." I raised my skinny arms to the blue sky and noticed them, and my ragged cuffs, and a mass of golden flowers that was hanging over me. Their color against the sky made my heart leap. Since then I have seen the same yellows, green, and blue in Van Gogh's harvest paintings and heard the same hurrahing in Hopkins's harvest poems, but my hurrahing, that made me inside out with exuberance, was for wildness. (Payne, 1998, pp. 38–39)

Payne's hurrah for wildness led her to a career studying the communication of whales and elephants and using that knowledge to better shape human interaction with these species. Having settled back into her home range around Ithaca after various personal and professional sojourns away, her present life and identity continue to be shaped by the intimate and energetic relationship with the natural world forged in her childhood.

Payne's experiences resonate in the lives and memories of many individuals, past and present, in the United States and across the globe. Reflecting this common experience, the study of childhood development has constituted one of the major avenues through which the natural world has entered into scholarly discussions of selfhood and identity. That this is so undoubtedly reveals the long shadow of William Wordsworth, whose Romantic formulation of the child's special capacities for perception and creativity in nature has infused Western culture for the past two centuries (Chawla, 1994, 2002).

In work beginning in the late 1950s, Edith Cobb (1977) found Wordsworth's basic conception of the child's creative interaction with the world (especially nature) echoed in the autobiographical reminiscences of a wide range of writers and artists. These often highlight the child's experience of "a revelatory sense of continuity—an immersion of his whole organism in the outer world of forms, colors, and motions in unparticularized time and space" (Cobb 1997, p. 88). Such experiences reflect "an aesthetic logic present in both nature's formative processes and the gestalt-making powers and sensibilities of the child's own developing nervous system. Inner and outer worlds are sensed as one in these moments of form-creating expansion and self-consciousness" (p. 110), providing grounds both for a childhood sense of identity and for the adult's capacities for artistic creativity. On the one hand, of course, this emphasis on an ideal unity of self and world based in childhood experience can (and should) be criticized as itself a cultural and historical construction, applicable to specific times, cultures, and groups of people, but by no means an objective description of universal reality. (As one element of such a critique, Chawla, 1990, suggests that Cobb's analysis is applicable only to the creative artists and writers who were her subjects, not to persons who ended up in other careers as adults.) On the other hand, the insight and power of these Wordsworthian beginnings have led psychological theorists and researchers to more nuanced explorations of childhood environmental experiences and the impact of those experiences on adult identity.

In perhaps the most sustained and insightful psychoanalytic treatment of the human relationship with nature, Harold Searles (1960) uses the theory of ego development through object relations to outline a process of development from "the infant's subjective oneness with his nonhuman environment" to a more mature sense of a relationship with the nonhuman realm. Searles proposes that for the young child, the "crucial phase of differentiation involves the infant's becoming aware of himself as differentiated not only from his human environment *but also from his nonhuman environment*" (1960, pp. 29–30; emphasis in original). In contrast to the Romantic perspective, for Searles the initial experience of infantile oneness is marked by the deep anxiety of association with what is at some times a "chaotically uncontrollable nonhuman environment." At other times during infancy, however, the world is experienced more positively—as "a harmonious extension of our world-embracing self"— and so the subsequent process of differentiation carries its own anxieties as well (1960, p. 39). Later, especially during adolescence, the natural world in particular provokes in the individual a "sense of inner *conflict* concerning his awareness that he is part of Nature and yet apart from all the rest of nonhuman Nature; and the two great ingredients of this inner conflict—man's *yearning to* become wholly at one with his nonhuman environment, and his contrasting *anxiety lest* he become so and thus lose his own unique humanness" (1960, p. 114; emphasis in original). Thus, negotiating this conflict, rather than either total rejection of nature or Romantic identification with it, opens the adult to a more mature and healthy adult selfhood and relationship to the world:

> It is my conviction . . . that the more directly we can relate ourselves to the non-human environment as it exists—the more our relatedness to it is freed from perceptual distortions in the form of projection, transference, and so on—the more truly meaningful, the more solidly emotionally satisfying, is our experience with this environment. Far from our finding it to be a state of negativity and deadness, we find in ourselves a sense of kinship toward it which is as alive as it is real. (Searles, 1960, p. 115)

Whatever one thinks of his specific convictions or conclusions, Searles establishes an important framework for discussing the affective, symbolic, and interpersonal dynamics of an individual relationship with nature. A number of more recent theorists have also developed the object relations approach in exploring the psychodynamics of environmental experiences from childhood onward, both with respect to place in general (Hart, 1979) and with special attention to the natural world (Holmes, 1999; Kidner, 2001).

Revisiting these themes with the help of theorists from Wordsworth through Jean Gebser, environmental psychologist Louise Chawla (1994) explores the persistence and use of childhood memories of nature in the adult consciousnesses of five contemporary American poets. In Chawla's analysis, these poets actively use and rework their memories of childhood to forge (or to transform) their adult self-identities and their poetic and personal visions of nature, with widely differing results—from joyous affirmation to defiant resignation. (For related work within more strictly empirical environmental psychology, see, e.g., Sebba, 1991; Daitch, Kweon, Larsen, Tyler, & Vining, 1996.)

Focusing on more specifically environmentalist research subjects and concerns, Chawla (1998, 1999) and others have researched the significant life experiences that contributed to the career paths and commitments of environmental professionals, with implications both for psychological research and for educational practice. Similarly, Thomashow's (1995) notion of "ecological identity" stands as both a focal point around which to discuss previous research and a guiding ideal or life-path clarification tool in environmental education. Indeed, Thomashow's book reminds us that most of the recent work on the developmental dimensions of the human relationship with nature has been undertaken in the context of awareness of ecological crisis and commitment to ethical, practical, and political responses. From whatever disciplinary location—even one valuing scientific research and objectivity—such work is always simultaneously an attempted description of individual development and an actual intervention in that development.

Place Theory

For many people, broad terms such as "the nonhuman environment" or "ecological identity" are too general, too abstract to have any meaning in their lives. What really matters to most people is not "the planet" as a whole, but rather specific *places—this* home, *this* soil, *this* town, area, or region. As one example, the lives and identities of the people living in the southern Appalachian area of the United States have historically and culturally been closely entwined with the hills and hollows that surround them. For one mountaineer interviewed in Tennessee in the 1960s, the possibility of having to leave his home place—which was forced upon many by economic circumstances—tore at his sense of self-worth and identity, a dilemma he placed before his son as well:

We're born to this land here, and it's no good when you leave. . . . But he knows what I'm telling him: for us it's a choice we have, between going away or else staying here and not seeing much money at all, but working on the land, like we know how to do, living here, where you can feel you're you, and no one else, and there isn't the next guy pushing on you and kicking you and calling you every bad name there is. (Coles, 1967, pp. 17–18)

"*Where you can feel you're you, and no one else.*" For this mountaineer, a sense of grounding in place is absolutely essential to his individual identity.

Similarly, African-American feminist bell hooks finds an important element of identity through a historical and communal continuity with place, but with greater options for carrying that identity with her to somewhere new:

As a child I loved playing in dirt, in that rich Kentucky soil, that was a source of life. Before I understood anything about the pain and exploitation of the southern system of sharecropping, I understood that grown-up black folks loved the land. I could stand with my grandfather Daddy Jerry and look out at fields of growing vegetables, tomatoes, corn, collards, and know that this was his hand-iwork. I could see the look of pride on his face as I expressed wonder and awe at the magic of growing things. I knew that my grandmother Baba's backyard garden would yield beans, sweet potatoes, cabbages, and yellow squash, that she too would walk with pride among the rows and rows of growing vegetables showing us what the earth will give when tended lovingly. (hooks, 1999, pp. 51–52)

The power of these memories challenges hooks to work to transplant that sense of place from Kentucky to New York City through her adult experiences of urban gardening: "I feel connected to my ancestors when I can put a meal on the table of food I grew" (1999, p. 56).

Our experience is never of "the earth" as an actual whole, but of some particular place on the earth, a place defined both by physical bound-aries and by the actions, concepts, meanings, and feelings that we enact within (or with) it—boundaries and behaviors that in turn play a role in defining us. Indeed, it is the specificity of place that allows it to serve as a basis for or reflection of individual identity; or perhaps place and self-hood are mutually codefining. In any case, a focus on place recognizes that some parts of the world are more important to an individual's iden-tity than other parts, or at least contribute to identity in different ways than others do—and sees this not as a limitation but as a source of strength. (For a history of the philosophical concept of place, see Casey, 1997.)

Although most formulations of place theory stress built and even imag-ined or symbolic dimensions as much or more than natural ones, such analysis is relevant to the importance of the physical environment for identity (both personal and social). In a sense, the entire field of human-istic geography takes this as its central concern: "We humans are geo-graphical beings transforming the earth and making it into a home, and that transformed world affects who we are" (Sack, 1997, p. 1). For many geographical theorists, place plays a central role in this process: "The mix of nature, meaning, and social relations that help constitute us are possible and accessible because of the activities of place and space.... The interthreading of place and self can thus oscillate from having places make us more aware of ourselves and our distinctiveness, to making us less aware, to the point where place and self are fused and conflated" (Sack, 1997, pp. 131, 136). Phenomenologically inclined geographers stress the constant creation of place as "world" or "life-world" through a bodily based subjectivity of perception, feeling, and action (e.g., Relph, 1976; Seamon, 1979).

Within environmental psychology, the intersection of self and place has been explored in a variety of ways, giving varying degrees of importance to the nonhuman world itself. For some theorists, place is psychologically meaningful primarily as a means of regulating human interactions through privacy, territoriality, and the personalization of space, with important implications for identity (e.g., Altman, 1975). Acknowledging the psychological and symbolic importance of the material world itself leads us closer to the notion of place identity. In one formulation, "place identity is conceived of as a substructure of the person's self-identity that is comprised of cognitions about the physical environment that also serve to define who the person is" (Proshansky & Fabian, 1987, pp. 22).

More recent work has proposed a deeper sense of relationship with the nonhuman world through attachment to place (Altman & Low, 1992). In particular, home—one of the most important places for humans, as for any species—has received special treatment from environmental psychologists (see, e.g., Altman & Werner, 1985) as well as from philosophers (Csikszentmihalyi & Rochberg-Halton, 1981; Bachelard, 1964) and design professionals (Marcus, 1995). Although none of these approaches specifically stresses the importance of the natural world, they all analyze the ways in which identity is shaped through experience of or relationship with particular nonhuman places and beings.

The Links among Identity, Ethics, and Action

For many, a sense of personal relationship to nature or place evokes an ethical commitment to practical action to protect and care for the natural environment—that is, environmental identity can lead directly to an identity as an environmental*ist*. Such identities are writ large in public figures such as John Muir and Rachel Carson. Muir's career as one of the founders of the turn-of-the-century conservation movement is indelibly associated in the public mind both with his beloved California mountains (especially Yosemite Valley) and with the Sierra Club, the environmentalist organization that he founded in 1894 to help protect those mountains. (See Holmes, 1999, for a theoretically informed analysis of the genesis of Muir's attachment to Yosemite, and, e.g., Turner, 1985, for the full story of how that attachment constituted the center of his subsequent environmental activism.)

In a similar manner, Carson proceeded from a fascination with and scientific study of the ecology of ocean and shore (which led to her initial literary fame and public recognition) to a broader environmental awareness: "I am not afraid of being thought a sentimentalist when I stand here tonight and tell you that I believe natural beauty has a necessary place in the spiritual development of any individual or any society" (Carson, 1998, p. 160). This sense of an ecologically grounded personal identity would lead her to a more specific public identity as well. Following the 1962 publication of *Silent Spring*, her groundbreaking exposé of the environmental and health dangers of pesticides, Carson became the archetypal modern environmentalist—the knowledgeable citizen whose understanding of the connection between environmental damage and human health leads to forceful action for governmental regulation of industrial pollution and corporate power.

The relationship between environmental identity and ethical action embodied in figures such as Carson and Muir has been analyzed philosophically in a variety of ways. Perhaps the most influential recent discourse relating identity and environmental ethics—in popular culture and environmental activism as well as in scholarly circles—has been that of deep ecology. While the dominant tradition in environmental philosophy is concerned with extending the concepts of rights or value to nonhuman beings (see, e.g., Nash, 1989), deep ecology explicitly calls for the extension of the sense of personal identity to include or encompass nature: "Spiritual growth, or unfolding, begins when we cease to understand or see ourselves as isolated and narrow competing egos and begin to identify with other humans from our family and friends to, eventually, our species. But the deep ecology sense of self requires a further maturity and growth, an identification which goes beyond humanity to include the nonhuman world" (Devall & Sessions, 1985, p. 67). The process of awakening to this ecological self includes both a personal dimension, "a humbling but also gratifying shift to a more expansive, accommodating, and joyous identity," and a more public and active aspect, as a "ground for effective engagement with the forces and pathologies that imperil us" (Macy, 1989, pp. 203, 202). Given its normative stance and concern for practical transformation, it is not surprising that deep ecology concerns itself both with issues of personal lifestyle (Devall, 1988) and with community building and ritual (Seed, Macy, Fleming, & Naess, 1988). Fundamentally, deep ecology is not so much a theory as a practice; to echo Marx, the real point is not so much to describe the relationship of self to the natural world, as to change it.

Of course, if understood simplistically, the basic deep ecology assertion of an expansion of selfhood to include nature can be absurd, vacuous, and/or arrogant, and so some deep ecologists have taken pains to develop the approach in more philosophically nuanced ways:

> What identification should not be taken to mean . . . is *identity*—that I literally *am* that tree over there, for example. What is being emphasized is . . . that through the process of identification my *sense* of self (my experiential self) can expand to include the tree even though I and the tree remain physically "separate" (even here, however, the word *separate* must not be taken too literally because ecology tells us that my physical self and the tree are physically *interlinked* in all sorts of ways). (Fox, 1990, pp. 231–232; emphasis in original)

The proposed fusion of person and planet should not be taken to deny the very real differences between the human and the nonhuman, or between various members of the nonhuman realm: according to one sympathetic feminist critic, "What is missing from deep ecology is a developed sense of *difference*. . . . A sense of oneness with the planet and all its life-forms is a necessary first step, but an *informed* sensibility is the prerequisite second step" (McFague, 1993, p. 128; emphasis in original).

In an effort to avoid such difficulties, other environmental philosophers propose a similar expansion of the sense of self, not through sheer "identification," but through a more active and relational approach; our emotions, relationships, actions, and intellect take us outside of ourselves and into contact with the world and it is (in part) through this contact that we come to be and to know who we are. In phenomenological terms, "the individual may profitably be thought of not as a thing but as a field," a range of integrated actions and emotions; in particular, "[i]f we were to regard ourselves as 'fields of care' rather than as discrete objects in a neutral environment, our understanding of our relationship to the world might be fundamentally transformed" (Evernden, 1985, pp. 43, 47; for other uses of phenomenology in environmental ethics, see, e.g., Abram, 1996; Clayton, 1998).

Moreover, it is within this tradition of philosophical and ethical speculation that I would locate the development in the 1990s of "ecopsychology." Ecopsychology's image of "a psyche the size of the earth"
echoes the expansion of self-identification proposed by deep ecology:
"Unlike other mainstream schools of psychology that limit themselves to
the intrapsychic mechanisms or to a narrow social range that may not
look beyond the family, ecopsychology proceeds from the assumption
that at its deepest level the psyche remains sympathetically bonded to
the Earth that mothered us into existence" (Roszak, Gomes, & Kanner,
1995, pp. xvii, 5). Despite the occasional theoretical insights and individual observations to be found in its writings, ecopsychology is not so
much a descriptive or empirical psychology as it is an ethical and practical outlook in response to the present environmental crisis; like deep
ecology, ecopsychology constitutes a self-transformative practice, and
indeed has been formulated as a therapeutic approach (e.g., Clinebell,
1996). At the same time, these normative philosophical perspectives are
by no means irrelevant for more conventional scholarly or scientific
approaches, and the packaging of ecopsychology as "psychology" may
help it serve as a moral guide or inspiration to the field of environmental psychology proper (Reser, 1995; Bragg, 1996).

Conclusion and Future Directions

This brief consideration of some lives and theories is not intended to be
a comprehensive survey of the issues but rather an illustrative and evocative tour through some of the intersections of environment and identity:
the primal importance of bodily or kinesthetic self-awareness in conditioning one's sense of identity; a sense of continuity across the life-span,
especially the integration of childhood memory in adult self-image; environmental experiences in the growth and maintenance of individuality
or uniqueness; self-definition and self-worth through assertion, work,
and achievement; the importance for identity of communal or regional
identity and of moral and political commitment; and what might be
called "ecological identity," or a felt relationship with natural beings,
places, and processes on their own terms (including, perhaps, "Nature"
or "the Earth" as an imagined or symbolic whole). On the one hand, all

of these elements are present to some extent in the life and identity of every individual; on the other hand, any one of them may take on a special importance in defining the shape and direction of a particular life.

Some of the important directions and challenges for future research in the field include increasing integration of interpretive approaches; recognizing the importance of culture in shaping environmental experiences; incorporating social and cultural diversity in research and analysis; and exploring a more inclusive range of occupations and activities than the usual focus on specifically environmentalist ones. I again take my cues from the life experiences of particular people in particular places.

Integration of Interpretive Approaches

For David Mas Masumoto, a third-generation Japanese-American peach farmer, personal, familial, and communal identity are all intermingled and rooted in place: "The greatest lesson I glean from my fields is that I cannot farm alone. . . . When I gaze over my farm I imagine Baachan [grandmother] or Dad walking through the fields. They seem content, at home on this land. My Sun Crest peaches are now part of the history of this place I too call home. I understand where I am because I know where I came from. I am homebound, forever linked to a piece of earth and the living creatures that reside here" (Masumoto, 1995, p. 229). Moreover, confronting the environmental and economic pressures of contemporary California agriculture, Masumoto's decision to "farm a new way, working with, and not against, nature" (p. 4) leads to transformations that are deeper than the merely agricultural. Among other things, Masumoto's personal and familial journey leads to a renewed commitment to low-impact, organic farming as a practical ethical expression of his love for the land—for *his* land, for that particular place. Thus, Masumoto's life (and life story) can be best understood through an integrated approach that incorporates all of the dimensions mentioned in the earlier literature review—psychological development, sense of place, and ethical action. Indeed, perhaps the most important lesson afforded by the study of personal life stories is the need to question and to cross disciplinary and intellectual boundaries in pursuit of more integrated, holistic, and encompassing perspectives on the complex reality of human lives and experience.

The Importance of Culture

As noted earlier, Katy Payne's rural childhood was filled not only with animals and open spaces but also with books and stories, both published and unpublished. In her adult work as a wildlife biologist, her ground-breaking investigations of elephant vocalization and communication were guided at important points not only by scientific research but also by her musical sensitivity and training (such as her childhood memories of feeling rather than hearing the low organ notes in Bach's *Passion According to St. Matthew*). Similarly, Rachel Carson notes the crucial influence of literature in shaping her early interest in the ocean, quoting Emily Dickinson:

I never saw a moor,
I never saw the sea,
Yet know I how the heather looks,
And what a wave must be.

Indeed, Carson "never saw the ocean until I went from college to the marine laboratories at Woods Hole. . . . Yet as a child I was fascinated by the thought of it. I dreamed about it and wondered what it would look like. I loved Swinburne and Masefield and all the other great sea poets" (1998, p. 148).

As these examples make clear, human experiences of nature—even in childhood—are never direct and unmediated; rather, perceptions are formed into experience and identity in part through the power of cultural symbols, ideas, and visions. This power has increased with the advent of mass media and electronic communication; for many young people, nature shows on television and stories of environmental destruction in the media and on the Internet may be as or more important than hands-on outdoor experience in forming an environmental consciousness. Thus, in fully understanding the genesis and shape of environmental attitudes and identity, we must attend to what people are reading, hearing, viewing, and fantasizing in school, at home, and in cultural settings, as well as to what they are doing and observing in the natural and built environments themselves. (For a broader statement on the cultural construction of nature, see Cronon, 1996.)

Diversity of Subjects

As some of the life stories mentioned here have already suggested, the role of the natural world in each person's life and identity is inextricably bound up with sociocultural factors such as gender, race, class, sexuality, occupation, ethnicity, and nationality—each factor historically contextualized in ways appropriate to the life of the subject in question. For example, in Borneo, the profession of "tree climber" has emerged in response to the needs of Western scientific expeditions for workers to identify and gather specimens from the high canopy of the rain forest (Primack, Goh, & Kalu, 2001). Master tree climbers such as Jugah Tagi and Banyeng Ludong gain both personal pride and public recognition—from scientific as well as local communities—through their skills in climbing high trees with minimal equipment and in identifying plant species at a glance; such work often involves incorporating both Western and indigenous systems of botanical knowledge, illustrating the blending of tradition and modernity in shaping the meanings of nature in individual lives. At the same time, the tree climbers' particular environmental identity also includes an element of resistance to the clash of cultures; their stories of using their superior knowledge and skill to save inexperienced Western scientists from the dangers of the rainforest, though told with characteristic humor, perhaps represent an assertion of self and culture in the face of the changes wrought by the invasion of the outside world.

For reasons of academic methodology and popular mythology, scholarly attention has traditionally been paid primarily to middle- and upper-class white persons (particularly males), ignoring the diversity of experience and perspective to be found in individuals and communities. Moreover, placing issues of gender, race, class, and political and power relationships at the center of our research and interpretation goes beyond the sheer inclusion of previously underrepresented groups (important as that inclusion is) to require a reshaping of the questions and categories of analysis overall. Not only the people of whom we ask the questions, but the questions themselves must take into account social location as both shaping and being shaped by environmental experience and identity. To note just one such axis of analysis, the example of the Borneo tree climbers suggests the ways in which the natural world can provide modes of identity and action from which dislocated or oppressed peoples can forge stances of power and resistance (even as the environmental justice movement emphasizes the increased risk of toxicity and environmental degradation borne by such groups). Toward the other end of the power spectrum, we must critically analyze the roles of gender, race, and class status in the lives and identities of the traditional environmental villains and heroes alike, from the Carnegies and Rockefellers to the Muirs and Carsons.

Inclusive Range of Occupations and Activities

Their engagement with nature defines the identities of working people and communities as well as academics and environmentalists (see also, e.g., White, 1996). For example, in the northern woods and mountains of the Adirondack region of New York State, anthropologist Katherine Henshaw Knott (1998) analyzes the "indigenous knowledge" possessed by loggers, guides, trappers, and maple syrup producers, exploring the implications of that knowledge for personal identity and social continuity. For Ross Putnam (one of Knott's interviewees), a lifelong relationship to and knowledge of the woods supports both his personal self-image and his role as father:

"I always liked the woods, even when we were small," he says. His boys used to go hunting with him all the time; he taught them all he could about survival—the quickest way to build a fire when it's wet, and the different kinds of trees. "Usually anything in the woods I notice, but I don't know a lot of the names. The bark on the yellow birch—you can eat it. It tastes just like wintergreen." (Knott, 1998, p. 145)

For another interviewee, Pierre LeBrun, a logger, the woods have shaped his selfhood in more bodily and sensuous ways: "In the woods is rough, but is good for yourself, for your health, the fresh air . . . I feel good. When you go into the woods, you change, you are not the same. It is so beautiful—the birds, the deer, the porcupine" (Knott, 1998, p. 114).

From a humanistic perspective, studying the environmental experiences and identities of workers and others—even those who participate in the destruction of the natural world—is valuable simply for acknowledging and exploring the widest range of human experience. At the same time, for purely environmentalist ends, it seems as crucial to learn how to stop destructive orientations as to promote conservationist ones; moreover, it may well not be from the political and social activists but rather from those quiet, unassuming people living with the land—farmers, workers, housekeepers, and others—that we have most to learn in forging rich and enduring patterns of identity and action within the natural world.

Building upon these and other intellectual and cultural resources, the scholarly approaches represented in this volume are important in formulating more deep, rich, and specific questions and insights about the intersections of identity and the natural world. However, the topic is much too important and elusive to be completely captured within any one body of scholarship, but rather is one of the common provocations, challenges, and joys of human intellectual and cultural life in general. Juxtaposing a variety of life stories, theories, and data, the work in this field possesses more than merely academic interest. It can help us all reach a deeper understanding of the interconnections of selves and the natural world, as a guide and inspiration toward the future health and flourishing of both.

References

Abram, D. (1996). *The spell of the sensuous: Perception and language in a more-than-human world*. New York: Pantheon.

Altman, I. (1975). *The environment and social behavior: Privacy, personal space, territory, crowding*. Monterey, Calif.: Brooks/Cole.

Altman, I., & Low, S. (Eds.). (1992). *Place attachment*. New York: Plenum.

Altman, I., & Werner, C. (Eds.). (1985). *Home environments*. New York: Plenum.

Bachelard, G. (1964). *The poetics of space*. New York: Orion.

Bragg, E. A. (1996). Toward ecological self: Deep ecology meets constructionist self-theory. *Journal of Environmental Psychology, 16*, 93–108.

Carson, R. (1998). *Lost woods: The discovered writing of Rachel Carson* (L. Lear, Ed.). Boston: Beacon.

Casey, E. S. (1997). *The fate of place: A philosophical history*. Berkeley: University of California Press.

Chawla, L. (1990). Ecstatic places. *Children's Environments Quarterly, 7*(4), 18–23.

Chawla, L. (1994). *In the first country of places: Nature, poetry, and childhood memory*. Albany: State University of New York Press.

Chawla, L. (1998). Significant life experiences revisited. *Journal of Environmental Education, 29*(3), 11–21.

Chawla, L. (1999). Life paths into effective environmental action. *Journal of Environmental Education, 31*(1), 15–26.

Chawla, L. (2002). Spots of time: Manifold ways of being in nature in childhood. In P. Kahn, Jr. & S. Kellert (Eds.), *Children and nature.* (pp. 199–225). Cambridge, Mass.: MIT Press.

Clayton, P. H. (1998). *Connection on the ice: Environmental ethics in theory and practice.* Philadelphia: Temple University Press.

Clinebell, H. J. (1996). *Ecotherapy: Healing ourselves, healing the earth: A guide to ecologically grounded personality theory, spirituality, therapy, and education.* Minneapolis: Fortress.

Cobb, E. (1977). *The ecology of imagination in childhood.* New York: Columbia University Press.

Coles, R. (1967). *Migrants, sharecroppers, mountaineers.* Boston: Little, Brown.

Cronon, W. (1996). "Introduction: In search of nature" and "The trouble with wilderness: or, Getting back to the wrong nature." In W. Cronon (Ed.), *Uncommon ground: Rethinking the human place in nature* (pp. 23–56, 69–90). New York: Norton.

Csikszentmihalyi, M., & Rochberg-Halton, E. (1981). *The meaning of things: Domestic symbols and the self.* Cambridge: Cambridge University Press.

Daitch, V., Kweon, B., Larsen, L., Tyler, E., & Vining, J. (1996). Personal environmental histories: Expressions of self and place. *Human Ecology Review, 3*(1), 19–31.

Devall, B. (1988). *Simple in means, rich in ends: Practicing deep ecology.* Salt Lake City, Utah: Peregrine Smith.

Devall, B., & Sessions, G. (1985). *Deep ecology: Living as if nature mattered.* Salt Lake City, Utah: G. M. Smith.

Evernden, N. (1985). *The natural alien: Humankind and environment.* Toronto: University of Toronto Press.

Fox, W. (1990). *Toward a transpersonal ecology: Developing new foundations for environmentalism.* Boston: Shambhala.

Hart, R. (1979). *Children's experience of place.* New York: Irvington.

Holmes, S. J. (1999). *The young John Muir: An environmental biography.* Madison: University of Wisconsin Press.

hooks, b. (1999). Touching the earth. In D. L. Barnhill (Ed.), *At home on the earth: Becoming native to our place: A multicultural anthology* (pp. 51–56). Berkeley: University of California Press.

Kidner, D. W. (2001). *Nature and psyche: Radical environmentalism and the politics of subjectivity.* Albany: State University of New York Press.

Knott, C. H. (1998). *Living with the Adirondack forest: Local perspectives on land use conflicts.* Ithaca, N.Y.: Cornell University Press.

Macy, J. (1989). Awakening to the ecological self. In J. Plant (Ed.), *Healing the wounds: The promise of ecofeminism* (pp. 102–211). Philadelphia: New Society.

Marcus, C. C. (1995). *House as a mirror of self: Exploring the deeper meaning of home.* Berkeley, Calif: Conari.

Masumoto, D. M. (1995). *Epitaph for a peach: Four seasons on my family farm.* San Francisco: HarperSanFrancisco.

McFague, S. (1993). *The body of God: An ecological theology.* Minneapolis: Fortress.

Nash, R. (1989). *The rights of nature: A history of environmental ethics.* Madison: University of Wisconsin Press.

Payne, K. (1998). *Silent thunder: In the presence of elephants.* New York: Simon & Schuster.

Primack, R., Goh, M., & Kalu, M. (2001). The view from the forest canopy. *Arnoldia 60*(4), 2–9.

Proshansky, H. M., & Fabian, A. K. (1987). The development of place identity in the child. In C. S. Weinstein & T. G. David (Eds.), *Spaces for children: The built environment and child development* (pp. 21–40). New York: Plenum.

Relph, E. C. (1976). *Place and placelessness.* London: Pion.

Reser, J. P. (1995). Whither environmental psychology? The transpersonal ecopsychology crossroads. *Journal of Environmental Psychology, 15,* 235–257.

Roszak, T., Gomes, M., & Kanner, A. (Eds.). (1995). *Ecopsychology: Restoring the earth, healing the mind.* San Francisco: Sierra Club Books.

Sack, R. D. (1997). *Homo geographicus: A framework for action, awareness, and moral concern.* Baltimore: Johns Hopkins University Press.

Seamon, D. (1979). *A geography of the lifeworld: Movement, rest, and encounter.* London: Croom Helm.

Searles, H. F. (1960). *The nonhuman environment, in normal development and schizophrenia.* New York: International Universities Press.

Sebba, R. (1991). The landscape of childhood: The reflection of childhood's environment in adult memories and in children's attitudes. *Environment and Behavior, 23,* 395–422.

Seed, J., Macy, J., Fleming, P., & Naess, A. (1988). *Thinking like a mountain: Towards a council of all beings.* Philadelphia: New Society.

Thomashow, M. (1995). *Ecological identity: Becoming a reflective environmentalist.* Cambridge, Mass.: MIT Press.

Turner, F. (1985). *Rediscovering America: John Muir in his time and ours.* San Francisco: Sierra Club Books.

White, R. (1996). 'Are you an environmentalist or do you work for a living?': Work and nature. In W. Cronon (Ed.), *Uncommon ground: Rethinking the human place in nature* (pp. 171–185). New York: Norton.

I
Experiencing Nature as Individuals

3

Environmental Identity: A Conceptual and an Operational Definition

Susan Clayton

Identity can be described as a way of organizing information about the self. Just as there are multiple ways of organizing this information, we have multiple identities, varying in salience and importance according to the immediate context and to our past experiences. Each level of identity may suggest its own perspective. When I am focusing on myself as unique I focus on issues that affect me personally and I may emphasize my personal merit and welfare, while when I think of myself as a group member I am aware of issues that affect my group, and the welfare of the group may be of primary concern. Understanding which identities are salient is important for understanding how people react to a particular threat or distribution of rewards.

Because the social aspects of identity are so obvious and so important, psychologists often overlook the impact of nonsocial (or at least non-human) objects in defining identity. Yet there are clearly many people for whom an important aspect of their identity lies in ties to the natural world: connections to specific natural objects such as pets, trees, mountain formations, or particular geographic locations (which has been studied under the rubric of "place identity"). This is not limited to those who are labeled "environmentalists" by virtue of a political position; many who are associated with positions that are considered antienvironmental nevertheless demonstrate, through words or behavior, their love of some aspect of the natural world.

This chapter develops the rationale for talking about an environmental identity. I propose that an environmental identity is one part of the way in which people form their self-concept: a sense of connection to some part of the nonhuman natural environment, based on history, emotional attachment, and/or similarity, that affects the ways in which we perceive and act toward the world; a belief that the environment is important to us and an important part of who we are. An environmental identity can be similar to another collective identity (such as a national or ethnic identity) in providing us with a sense of connection, of being part of a larger whole, and with a recognition of similarity between ourselves and others. Also like a group identity, an environmental identity can vary in both definition and importance among individuals. I present a scale for assessing individual differences in environmental identity and evidence for the utility of such a scale in predicting reactions to environmental issues. I close by discussing the implications of environmental identity for personal and political behavior.

Self-concept can vary from situation to situation; William James argued that people can be said to have as many social selves as there are groups of people whose opinions matter to them (James, 1950). Why single out environmental identity as particularly worthy of examination? An identity is both a product and a force (see Rosenberg, 1981): an assortment of beliefs about the self and a motivator of particular ways of interacting with the world. In this chapter I argue that the natural environment has the potential to be a distinctively rich source of self-relevant beliefs.

As a motivating force, a strong environmental identity also can have a significant impact by guiding personal, social, and political behavior. (Chapter 2 gives some examples of people whose environmental identity has guided their life choices.) The current importance of, and concern about, environmental behavior thus provides another reason for examining environmental identity.

Environmental Identity as Product

Although the impossibility of experimental research precludes a definite conclusion, we can speculate that environmental identities come from interactions (e.g., with the natural world) and from socially constructed understandings of oneself and others (including nature) (Chawla, 1999). In order to conclude that these experiences have resulted in a significant identity, we should find that they are emotionally significant and that they affect the ways in which people think about themselves.

Emotional Connections to the Natural Environment

People tend to value nature, as evinced by their willingness to expend resources of time or money on nature (paying a premium in order to have natural views out one's window; devoting many hours to one's garden) or to endure discomfort (sleeping on the ground, hiking or driving long distances) in order to interact with it. This is not to deny that some people have lower thresholds for such expenditures or discomfort than others (see, e.g., Bixler & Floyd, 1997), but for most people there is at least some value to experiences with the natural environment that cannot be accounted for by the usual motivators of money, sex, food, and social status. Consider the following statistics for the United States: In 2001, 59 percent of a representative sample of the population thought that protecting the environment was more important than producing energy (CBS News poll, September 4, 2001); 52 percent of a different sample said protecting the environment was more important than encouraging economic growth (ABC News poll, August 1, 2001), and 61 percent of a third sample agreed that "protecting the environment is so important that requirements and standards cannot be too high and continuing environmental improvements must be made regardless of cost" (CBS/*New York Times* poll, March 13, 2001).

Systematic empirical research has shown fairly consistently that people prefer natural to human-influenced settings. They give high ratings to slides or photographs of natural scenery compared with scenes of the built environment (Herzog, Black, Fountain, & Knotts, 1997); among scenes of the built environment, those with some natural component (trees, flowers) are preferred to those without them; even undistinguished natural scenes are preferred to fairly attractive urban scenes or townscapes (R. Kaplan & Kaplan, 1989). Even in guided imagery, natural scenes are preferred over non-natural scenes (Segal, 1999). When people are asked to describe a favorite place, natural settings predominate (Korpela, Hartig, Kaiser, & Fuhrer, 2001; Moore, 1990). Nature doesn't have to be "out there," at a remove from daily living, in order to be valued. In an extensive investigation of "the meaning of things" for individuals, Csikszentmihalyi and Rochberg-Halton (1981) asked members of eighty-two families what things in their home were most special to them. Fifteen percent of the respondents (forty-seven people) mentioned plants; significantly, plants were described as embodying personal values more often than any other type of object.

Public support for policies that protect the environment tends to be quite high (Ritov & Kahneman, 1997), so much so that even ads for products and businesses that are harmful to the environment are often portrayed as promoting environmental protection, a phenomenon known as "greenwashing" (Bruno, 2001). There are some differences in support that are related to individual background, but they do not cancel out the general tendency. Despite stereotypes to the contrary, an appreciation for the natural environment does not seem to be limited by ethnicity or economic status (see, e.g., Kahn, 1999; Morrissey & Manning, 2000).

One explanation for the attraction of natural settings is the biophilia hypothesis, first proposed by E. O. Wilson (1984) and elaborated since then by Kellert and Wilson (1993). The biophilia hypothesis proposes that humans have inherited a genetic tendency to respond to the natural environment in certain ways, particularly with certain emotional responses. Such a tendency, it is argued, would have been adaptive in a hunter-gatherer type of society, leading us to attend to important environmental dangers, such as tigers and snakes, with a high degree of interest and to prefer to live in settings that would be more conducive to our survival. Like all theories about a genetic basis for human behavior, it is difficult to prove and controversial, although there is some evidence to support it (see Ulrich, 1993). If the hypothesis is valid, then a connection to nature is a fundamental part of who we are.

Without referring to prehistory or genetics, the preference for natural environments may be explained as being due to their psychological and physiological benefits. Kellert (1997) has detailed the many direct and indirect benefits we derive from nature even today, including improved physical fitness, self-confidence, curiousity, and calm. Positive effects of nature have been found on students, the elderly, and hospital patients, as measured by self-report, performance, physiological indices of arousal, and recordings of brain activity. Exposure to a view of a natural environment (e.g., through a window) or access to a nearby natural setting such as a park or garden, has been shown to strengthen cognitive awareness, memory, and general well-being and to decrease depression, boredom, loneliness, anxiety, and stress (Hartig, Mang, & Evans,

1991; Parsons, Tassinary, Ulrich, Hebl, & Grossman-Alexander, 1998; Ulrich, 1984).

R. Kaplan and Kaplan (1989) and S. Kaplan (1995) have discussed the ability of the natural environment to serve as a restorative setting for people. By engaging our attention, taking us away from our mundane activities and settings, and providing a level of stimulation that is neither overly arousing nor boring, the natural environment allows us to replenish our attentional resources. Research by the Kaplans and others (Herzog et al., 1997; Korpela et al., 2001) has shown that people rate natural settings as the best place for achieving these restorative goals.

Environment and Self-Knowledge

To constitute an important aspect of identity, the natural environment must influence the way in which people think about themselves. The natural environment provides the opportunity not only for attentional restoration but also for self-reflection (Herzog et al., 1997). By allowing people the time and space to think about their own values, goals, and priorities, as well as, perhaps, providing relief from the usual concerns of self-presentation, the natural environment can play a vital role in the extent to which we define ourselves to ourselves. One study (Young & Crandall, 1984) even found a slight positive relationship between wilderness use and self-actualization. Self-actualized people (e.g., Maslow, 1970) are described as more accepting of themselves, more sensitive to their own feelings, more independent of their social environment, and as fulfilling their own potential: "They are people who have developed or are developing to the full stature of which they are capable" (Maslow, 1970, p. 150).

The self is often described as including the knower and the known—that is, that which perceives the self and the content of what is perceived (see Ashmore & Jussim, 1997). The natural environment may enable a person to become both a more perceptive knower and a more positively valued known. We understand ourselves better and like ourselves more.

Nature can provide increased understanding of our own abilities and influence in part because it does not change very much in response to a person's behavior; only our own position in nature changes (Scherl, 1989). In a social environment, other people respond to us in a way that is affected, not only by our own behavior, but also by their perception of our behavior, our appearance and status, and by other forces that are invisible to us. If someone is rude to me, I cannot always tell what I may have done to elicit such behavior, or if I have done anything at all. This makes it difficult to predict the effect or outcome of my behavioral choices in the future. In a natural environment, it is clear what can be controlled and what cannot. If I am cold, it is because I did not accurately anticipate or appropriately dress for the local weather. The weather changes, but not in response to my behavior. Animals often do respond to us, but their motives (self-protection, hunger) tend to be straightforward and the impetus for their behavior clear. The link between my behavior and its consequences may be clearer in a natural than in a social environment.

In a broader sense, the natural environment informs us about what it means to be human (see chapter 4). We can only gain a thorough sense of our human identity—including, perhaps, an appropriate level of humility in the face of our own limitations—through comparison with nonhuman entities.

The natural environment also may be able to encourage a strong and positive sense of self. Three qualities seem to be desired parts of everyone's identity: autonomy, or self-direction; relatedness, or connection; and competence (Ryan & Deci, 2000). Perceived autonomy can be enhanced in a natural environment because there are fewer commands or requests from others that limit behavioral choices. Although the physical environment provides some constraints on behavior, the limits are less explicit than in a social setting where one is ruled by laws, signs, and the expectations of others. Fredrickson and Anderson (1999) obtained detailed information about the experiences of women on trips in the wilderness. On the basis of interviews and journals, Fredrickson and Anderson state "For many of these women what made their wilderness experience especially meaningful was the fact that there was virtually no reason to be anyone but themselves" (1999, p. 30).

Relatedness comes from the opportunity to feel like a part of a functioning system. Aron and McLaughlin-Volpe (2001) have argued that "self-expansion" is a fundamental human motivation, which can be achieved through redefining the self in a way that includes others. Many seem to reach this state through experiences with the natural environment. For some people it is experienced as a spiritual dimension, a connection with Mother Earth, or a unity with Gaia; for others, it is simply a way of fitting oneself into the larger picture, as part of an environment, a world, a functioning ecosystem. This redefinition of self is one of the primary tenets of deep ecology (Naess, 1989).

Competence comes from the feeling of self-sufficiency: getting around under our own steam; perhaps showing survival skills through living off the land, climbing a steep mountain, or staring down one's fear of the dark. The women interviewed by Fredrickson and Anderson (1999) talked about the sense of accomplishment they experienced, and of getting to know their own physical capabilities. R. Kaplan and Kaplan (1989) have shown that people's preference of natural to human-influenced environments is related to their perceived competence in a natural environment. The sense of competence can also be a significant part of experiences with nature in an urban setting (see chapter 10). It is interesting that De Young (2000) has also emphasized the importance of the desire for competence in motivating environmentally sustainable behavior.

The natural environment thus seems to provide a particularly good source of self-definition, based on an identity formed through interaction with the natural world and on self-knowledge obtained in an environmental context.

Environmental Identity as Force

The validity and utility of the construct of environmental identity depend on some evidence that it affects our thinking or behavior, and does so better than other determinants, such as attitudes. The idea of environmental identity enables us to see the relevance of the self to interactions over environmental issues. Environmental attitudes are espoused not, in many cases, because of a belief that they will be listened to, but in order to define oneself by expressing one's fundamental values and ethical standards (L. L. Thompson & Gonzales, 1997). Donations to environmental organizations are not made to increase perceived personal utility but to obtain moral satisfaction (Ritov & Kahneman, 1997). Environmental behaviors are engaged in, or not, in part because of the way in which these behaviors convey information about the self and group affiliations.

This raises the issue of individual differences in the strength and/or definition of an environmental identity. Some people receive validation and pleasure from reducing resource use; as De Young (2000) has noted, environmentally sustainable behavior becomes self-interested when an individual cares about the natural environment and values its well-being. Clearly, however, there are others who do not find environmental behaviors intrinsically rewarding.

Schultz (2000) has argued that environmental concerns are a function of the extent to which "people include nature in their cognitive representation of self" (p. 403). Previous research (Davis, Conklin, Smith, & Luce, 1996) has shown that a perspective taking manipulation increases the perceived interconnectedness between the self and others. Schultz found support for his hypothesis when he encouraged college students to adopt the perspective of an animal that was shown as being harmed by environmental degradation (oil spills, litter, etc.), thus presumably increasing the perceived relatedness between the student and the animal. This manipulation led to an increase in the students' rated concerns for environmental entities, although merely viewing the picture without taking the animals' perspective did not have that effect. In related work, Opotow (e.g., 1994; Opotow & Weiss, 2000) has discussed how concern for environmental entities is based on including those entities within our moral community.

The Environmental Identity Scale

In order to examine whether individual differences in environmental identity can predict behavior, I have constructed a scale to assess the extent to which the natural environment plays an important part in a person's self-definition—what I have called an Environmental Identity (EID) Scale (see the appendix). The structure of the scale was based in part on discussions of the factors that determine a collective social identity, such as an ethnic identity (cf. Luhtanen & Crocker, 1992; Sellers, Smith, Shelton, Rowley, & Chavous, 1998; Tajfel, 1981). Such factors include the salience of the identity, the identification of oneself as a group member, agreement with an ideology associated with the group, and the positive emotions associated with the collective.

With regard to the natural world, I operationalized salience by asking about the extent and importance of an individual's interactions with nature (e.g., "I spend a lot of time in natural settings"). Self-identification was assessed through the way in which nature contributes to the collectives with which one identifies ("I think of myself as a part of nature, not separate from it"). Ideology was measured by support for environmental education and a sustainable lifestyle. ("Behaving responsibly toward the Earth—living a sustainable lifestyle—is part of my moral code.") Positive emotions were measured by asking about the enjoyment obtained in nature, through satisfaction and aesthetic appreciation ("I would rather live in a small room or house with a nice view than a bigger room or house with a view of other buildings"). On the assumption that environmental identity results from experiences with nature, the scale also includes an autobiographical component, based on memories of interacting with nature ("I spent a lot of my childhood playing outside").

It is important to acknowledge that an environmental identity is also at least in part a social identity (see chapter 14). An understanding of oneself in a natural environment cannot be fully separated from the social meanings given to nature and to environmental issues, which will vary according to culture, world view, and religion. The scale I have created is very much mediated by North American understandings of the ways in which we value and interact with nature, and research with the EID scale has thus far been limited to North American college students. I would expect some, but not all, of the constructs to vary across cultures.

In several studies that have been conducted using the EID scale, it has proven to have good internal reliability. In three studies, Cronbach's alpha has been 0.90 or above for the twenty-four–item scale. Factor analyses suggest that a single factor accounts for most of the variance, although this may change when the test is administered to a more diverse sample. Although social desirability pressures might be expected to lead to high scores, the scale is relatively free from floor or ceiling effects; out of a possible range of 24–196, it had an actual range of 62–191 and a mean of 128.4 in one study and 63–180 and a mean of 125.4 in another. Women tend to score slightly higher than men, but the differences are usually not significant.

Study 1: The Validity of the EID Scale

To evaluate the convergent, discriminant, and predictive validity of the EID scale, I had seventy-three students complete a series of related measures assessing environmental attitudes, general values and ideology, and behavior.

Identity Compared with Attitudes and Values   The first measure was the Environmental Attitudes Scale (S. Thompson & Barton, 1994). This measures three components of environmental attitudes: ecocentrism, or a valuing of nature for its own sake; anthropocentrism, or a valuing of nature for its utility to humans; and apathy toward the environment and environmental issues.

The students also completed a twenty-four–item value survey, similar to that described by Schwartz (1992), in which they rated twenty-four values according to the importance of the value in their own lives. Three values were included to tap each of eight dimensions identified by Schwartz: self-enhancement, hedonism, stimulation, universalism, self-determination, benevolence, tradition, and conformity. The "universalism" items were specifically selected to represent environmental values ("a world of beauty," "protecting the environment," and "unity with nature"), while other items that Schwartz classified as universalist were considered to fall in the "benevolence" category (equality, social justice). A factor analysis confirmed that eight essentially uncorrelated factors were represented, and that the environmental items loaded primarily on a different factor than did the benevolence items.

A third instrument was designed to measure individualism and collectivism, and was abbreviated from one suggested by Triandis (1995). This was designed to assess the extent to which participants think of themselves primarily as independent individuals or as interdependent members of a larger community. Triandis summarizes the differences between an individualistic world view and a collectivist world view as follows: Collectivists think of the self as interdependent, believe that personal and communal goals are aligned, focus on their interpersonal norms and obligations, and emphasize relationships. Individualists think of the self as independent of others, believe that personal and communal goals are not the same, focus on their personal rights and attitudes, and evaluate relationships in terms of personal utility. Triandis also makes a distinction between a vertical and a horizontal perspective, either of which can accompany individualism or collectivism. A vertical perspective emphasizes difference and hierarchy, while a horizontal perspective emphasizes similarity. In the present study, only horizontal collectivism and vertical individualism emerged as coherent subscales.

I have argued elsewhere (e.g., Clayton, 1996, 1998) that an environmental perspective is more compatible with collectivism than with individualism. Environmentalists by definition tend to focus on the larger community—the species or the ecosystem. Environmental arguments stress interdependence and often refer to duties and obligations. The position opposed to the environmental argument, in contrast, focuses on individual entitlements (e.g., property rights) and often relies on market pricing to resolve environmental conflicts—a classic individualist approach (Opotow & Clayton, 1998).

I expected the EID scores to correlate positively with ecocentrism and negatively with apathy, to correlate with a high rating for the universalism factor from the value survey, and to correlate positively with horizontal collectivism and negatively with vertical individualism. In all cases, the EID score showed a pattern of correlations consistent with the hypotheses. It was significantly correlated with ecocentrism (0.79), apathy (−0.69), the universal values factor (0.66) but no other values, horizontal collectivism (0.37), and vertical individualism (−0.29).

Identity and Behavior   The last and probably most important measure to show the validity of the EID scale was a measure of behavior. In the same study, I devised a twenty-one-item self-report of behavior, which asked students to indicate on a five-point scale the extent to which they engaged in a range of environmentally sustainable actions, from "turn lights off when leaving a room" (this item had the highest mean, at 4.18), to "donate money to environmental organizations" (this had the lowest mean, at 1.58). The overall average mean was 2.82, suggesting that students were not overly inflating their estimates of reported actions and that social desirability pressures were not creating a ceiling effect. The overall internal reliability for this scale was 0.85. The EID scores showed a significant correlation with environmental behaviors, $r = 0.64$.

It could be argued that the EID measure is essentially replicating one of the other measures, most likely environmental attitudes, but possibly also values or collectivist ideology. To test this possibility, I calculated two sets of partial correlations: the relationship between the EID score and behavior while holding constant the other predictor variable (attitudes, values, or ideology), and the relationship between the other predictor variable and behavior while holding the EID score constant. In all cases, the correlation between the EID score and behavior remained significant even with the other variable held constant, but the correlation between the other variable and behavior became insignificant when the EID score was held constant. This shows that the relationship between the EID score and behavior is due primarily to a unique variance not shared with the other predictors, but that the relationship between the other predictors and behavior is due to the variance shared with the EID score.

Study 2: Identity and Environmental Conflict

The EID scale seemed to have personal relevance as a predictor of individual behavior. I also wanted to determine whether it might predict a more social outcome by affecting position in an environmental conflict. Eighty college students were presented with written descriptions of two environmental conflicts, which were adapted from newspaper accounts of real environmental issues. The students were presented with two possible resolutions of the conflict (one that would protect the environment and one that would harm the environment) and asked to decide which they preferred. They also rated the perceived importance of their decision, the difficulty of making the decision, and their certainty that they had made the right decision. Finally, they completed the EID scale.

Clearly, college students reading about an issue may make different decisions than people in an actual conflict situation; these choices should not be interpreted as predicting the corresponding behavior. However, many environmental attitudes are formed on the basis of exposure to just such incomplete information, and college students have as much exposure to environmental issues as most other citizens. The choices could represent political choices, such as votes for or against a policy.

The EID score was significantly related to individuals' decisions, so that a higher score was associated with the proenvironmental choice ($r = 0.27, 0.38$ for the two conflicts); to how important they thought the decision was ($r = 0.45, 0.38$); to how certain they felt that their decision was right ($r = 0.35, 0.20$); and interestingly, to a tendency to find the decision less difficult (although this did not quite reach standard significance levels; $r = -0.21, -0.20$). Previous research has shown that we process information that is self-relevant more quickly than information that is not, which may explain the correlation with difficulty (Markus, 1977); more research is needed to substantiate this hypothesis.

Study 3: Identity and Justice

A third study was conducted to examine whether the EID score related not just to the decision a person makes but also to the criteria for a good decision. It could be argued that a decision one makes about a public conflict is essentially a political decision, and could be indicative of a number of different motivations, including the sense that a particular position is more "politically correct" or socially desirable. (In the second study, 67 of 80 made a proenvironmental decision in one scenario and 53 of 80 chose the proenvironmental position in the other scenario.) The principles upon which one relies to make a good decision, however, should be tied to deeper personal beliefs.

In this study, 115 students were presented with three different environmental conflict scenarios: one describing a conflict between city governments and low-income city residents over abandoned lots that were developed into community gardens, one about the conflict between making national parks accessible to the public and leaving them undeveloped, and one concerning government regulation of private land development. Following each scenario, the students were asked not to resolve the conflict, but to indicate how much weight they would give to each of thirteen different factors in resolving the conflict. These factors were chosen to represent different principles of justice that could be considered in trying to arrive at a fair decision, including equality, equity, fair process, rights, and responsibilities as applied to different moral entities. The order of presentation was randomized. Following these ratings, students filled out the EID form.

The EID score was generally unrelated to nine of the thirteen principles, but was consistently (across scenario) positively related to the rated importance of two factors: "responsibility to other species," and "the rights of the environment," (multivariate $F [3, 108] = 7.64$, $p < .001$; and 7.98, $p < .001$, respectively; univariate F's for each scenario were also significant). A third principle, "responsibility to future generations," was related to the EID score in each scenario at levels that approached significance ($p < .10$). In other words, people high in environmental identity accord more weight than people low in environmental identity to those principles that endow environmental entities with moral standing. The EID score was also related to an increased rating for a fourth principle, "managing natural resources for the public good," in the abandoned lots and private lands scenarios (and achieved overall significance: multivariate F [3, 107] = 3.79, $p < .05$), but was unrelated to this principle in the national parks scenario, probably because the interests of the public and the interests of the natural environment were potentially conflicting in that case.

In sum, environmental identity seems to serve as a part of individuals' self-definition and, as such, as a predictor of individual behavior. It also has significance as a determinant of how individuals position themselves with regard to social situations, such as environmental policies or conflicts over environmental resources. Not just position, but also valued principles and certainty about one's position are related to environmental identity.

The Uses and Limits of Identity in Understanding Political and Personal Behavior

The Environmental Identity Scale is presented as a useful tool for measuring individual differences in the sense of the self in relation to the natural environment in order to draw conclusions about other variables associated with those differences. As such, it appears to have good psychometric properties. The research presented here has supported the idea that environmental identity is a meaningful and measurable construct, with consequences for attitudes and behavior, and that by thinking about environmental identity we learn something beyond what we learn by talking about attitudes and values. The EID scale is not intended to be a definitive operationalization of environmental identity. As is clear from this book as a whole, there are myriad ways of thinking and talking about environmental identity, most of which are not mutually exclusive.

The EID scale is also presented in the hopes of furthering consideration and stimulating research on environmental identity as a useful psychological construct. What does it mean to think about individuals as having varying levels of an environmental identity? What questions can usefully be empirically addressed? The EID scale, administered before and after an environmental intervention, might serve as a useful measure of the extent to which real-world policies have affected the way in which individuals think about their relationship to the environment.

Identities originate within a social context that gives meaning to our encounters with nature. Identities also have social significance, promoting certain group affiliations and activities and discouraging others. The focus of this chapter is primarily on individual identity as a product of individual experiences; the social implications of identity, however, are clear. In many ways, the natural environment has become a political issue, with highly controversial policies constantly on the public agenda. Environmental identity can function to affiliate people with well-defined social groups. It can become a political identity in a way that is divisive rather than integrative. In trying to understand people's reactions to environmental issues, we need to understand that positions are taken and behaviors engaged in, not just because of an assessment of costs and benefits, but partly because of the associations between these positions and behaviors and group affiliations. There are bumper stickers that contain slogans like "pave the rainforest" and "nuke the baby seals." It seems unlikely that the drivers whose cars are decorated by these stickers really have anything against the rainforest or baby seals; rather, they are hostile toward the people ("environmentalists") whom they associate with support for the rainforest and for baby seals. Environmental politics has become identity politics, and conflicts can escalate as people link their identity to particular positions (see chapter 12). It would be instructive to administer the EID scale to groups on opposite sides of an environmental conflict and see if changes occur as positions harden and the groups polarize.

Compared with a group identity, an environmental identity has some significant differences. It has the advantage of having no fixed geographic boundaries; this means that it is difficult to tie it to another group classification, such as a national identity. The protean membership of groups defined from an environmental identity perspective should make it harder for intergroup conflicts to become firm, as has been discovered in alliances between hunters and vegetarians to protect wildlife habitat, or between labor and environmental groups to reduce exposure to hazardous waste. Affiliations with the natural environment may thus be used to overcome intergroup hostility, as with the cross-national "peace parks" that are designed to encourage neighboring countries to work together to protect an ecosystem that does not stop at national boundaries (Weed, 1994). Similarly, Oskamp (2000) suggests that environmental sustainability can be used as a superordinate goal toward which all nations and peoples can work, overcoming their internal hostilities. Perhaps if the EID scale were used with opposing groups, we might discover that despite group differences on some items, there were similarities on others. These could serve as a foundation to encourage the recognition of intergroup similarities.

Moving beyond the use of the scale, we can use the idea of an environmental identity to motivate environmental behavior. An environmental identity locates us within a collective that is truly an interdependent system. If we recognize the significance and value of other members of the system, including nonhuman entities, that is one step toward acknowledging the (limited) rights of those entities, the way in which they are affected by our own actions, and the obligations that we owe them (Clayton, 1996). Cause and effect, the role that people play in maintaining the health of the system, are more easily seen than in a social group. An environmental identity can be recognized, nurtured, and used to encourage conservation behavior when the natural objects being protected are tied to the self, thus allowing the motivation to be internal rather than external. This is why people "buy" an acre of the rain forest for their very own (and receive a certificate to that effect), or "adopt" a particular whale, seal, or stretch of highway. When the health and success of a natural object are made more self-relevant, they are worth more to the individual.

This recognition of interdependence and obligation, however, also makes it hard to acknowledge an environmental identity. Acknowledging an environmental identity entails a shift in worldview that may be difficult (see, e.g., Dunlap & Van Liere, 1978; Dunlap, Van Liere, Mertig, & Jones, 2000). It removes us from the center; the value of things is not based only on their value to us. It limits our control; we have to love what we get rather than create what we want. It forces us to really recognize and value difference rather than ignoring it. A response can be to deny environmental identity or to anthropomorphize (see chapter 5)—to pretend animals and even trees are just like us and thus can thrive in a human-dominated world.

The fact that we have different sources of identity means that an environmental identity is not enough to ensure environmental responsibility (see chapter 6). An environmental identity has to be salient, and it has to motivate us to gain the knowledge we need to understand our impact on the environment. It is easy to stop with our own local or personal bit of nature: to support a neighborhood park but not the preservation of national parks; to support zoos but not protect the habitat of animals we will never see or do not like; to plant a garden but condone the destruction of rainforests. We are all challenged to walk the line between denying connection and denying difference: learning to accept responsibility without ownership. Environmental identity as assessed by the EID scale is perhaps less about differences in emotional connection to nature and more about differences in the significance, prominence, and scope of that connection. To give a twist to a phrase from the women's movement, an environmental identity is what we need in order to recognize that the personal is political and vice versa: that our immediate local actions can have global consequences, and that remote environmental threats are personally significant.

When I go to my local garden center, especially on a sunny day in spring, there are large numbers of people spending hard-earned money on flowers, vegetables, shrubs, trees, and grass seed. Their motives, although diverse, are almost certainly self-relevant. They enjoy the feeling of competence derived from digging in the dirt and getting things to grow; or they may be concerned with self-presentation, wanting their garden to be admired by the neighborhood. Their behavior is not necessarily proenvironmental; on the contrary, they may be harming the local environment by applying pesticides, replacing native with nonnative species, and breaking up the habitat of area wildlife. Rather than feeling pessimistic about their behavior, though, I am encouraged by the underlying value it reflects. The connection to nature is there.

Appendix: Items on the Environmental Identity Scale

———1. I spend a lot of time in natural settings (woods, mountains, desert, lakes, ocean).
———2. Engaging in environmental behaviors is important to me.
———3. I think of myself as a part of nature, not separate from it.
———4. If I had enough time or money, I would certainly devote some of it to working for environmental causes.
———5. When I am upset or stressed, I can feel better by spending some time outdoors "communing with nature."

————6. Living near wildlife is important to me; I would not want to live in a city all the time.

————7. I have a lot in common with environmentalists as a group.

————8. I believe that some of today's social problems could be cured by returning to a more rural life-style in which people live in harmony with the land.

————9. I feel that I have a lot in common with other species.

————10. I like to garden.

————11. Being a part of the ecosystem is an important part of who I am.

————12. I feel that I have roots to a particular geographic location that had a significant impact on my development.

————13. Behaving responsibly toward the Earth—living a sustainable life-style—is part of my moral code.

————14. Learning about the natural world should be an important part of every child's upbringing.

————15. In general, being part of the natural world is an important part of my self-image.

————16. I would rather live in a small room or house with a nice view than a bigger room or house with a view of other buildings.

————17. I really enjoy camping and hiking outdoors.

————18. Sometimes I feel like parts of nature—certain trees, or storms, or mountains—have a personality of their own.

————19. I would feel that an important part of my life was missing if I was not able to get out and enjoy nature from time to time.

————20. I take pride in the fact that I could survive outdoors on my own for a few days.

————21. I have never seen a work of art that is as beautiful as a work of nature, like a sunset or a mountain range.

————22. My own interests usually seem to coincide with the position advocated by environmentalists.

————23. I feel that I receive spiritual sustenance from experiences with nature.

————24. I keep mementos from the outdoors in my room, such as shells or rocks or feathers.

References

ABC News poll (August 1, 2001). Retrieved March 25, 2003 from Lexis Nexis.

Aron, A., & McLaughlin-Volpe, T. (2001). Including others in the self: Extensions to own and partner's group memberships. In C. Sedikides & M. B. Brewer (Eds.), *Individual self, relational self, collective self* (pp. 89–108). Philadelphia: Psychology Press.

Ashmore, R. D., & Jussim, L. (1997). Introduction: Toward a second century of the scientific analysis of self and identity. In R. D. Ashmore & L. Jussim (Eds.), *Self and identity: Fundamental issues* (pp. 3–19). New York: Oxford.

Bixler, R. D., & Floyd, M. F. (1997). Nature is scary, disgusting, and uncomfortable. *Environment and Behavior, 29*, 443–467.

Bruno, K. (2001). Greenwash, Inc. *Sierra* [the magazine of the Sierra Club], May/June, pp. 82–83.

CBS News poll (Sept. 4, 2001). Retrieved March 25, 2003 from Lexis Nexis.

CBS/ *New York Times* poll (Mar. 13, 2001). Available from Roper Center Public Opinion online.

Chawla, L. (1999). Life paths into effective environmental action. *Journal of Environmental Education, 31*, 15–26.

Clayton, S. (1996). What is fair in the environmental debate? In L. Montada & M. J. Lerner (Eds.), *Current societal concerns about justice* (pp. 195–211). New York: Plenum.

Clayton, S. (1998). Preference for macrojustice versus microjustice in environmental decisions. *Environment and Behavior, 30*, 162–183.

Csikszentmihalyi, M., & Rochberg-Halton, E. (1981). *The meaning of things.* New York: Cambridge.

Davis, M. S., Conklin, L., Smith, A., & Luce, C. (1996). Effect of perspective taking on the cognitive representation of persons: A merging of self and other. *Journal of Personality and Social Psychology, 70*, 713–726.

De Young, R. (2000). Expanding and evaluating motives for environmentally responsible behavior. *Journal of Social Issues, 56*(3), 509–526.

Dunlap, R. E., & Van Liere, K. D. (1978). The "new environmental paradigm": A proposed measuring instrument and preliminary results. *Journal of Environmental Education, 9*, 10–19.

Dunlap, R. E., Van Liere, K. D., Mertig, A. G., & Jones, R. E. (2000). Measuring endorsement of the New Ecological Paradigm: A revised NEP scale. *Journal of Social Issues, 56*, 425–442.

Fredrickson, L. M., & Anderson, D. H. (1999). A qualitative exploration of the wilderness experience as a source of spiritual inspiration. *Journal of Environmental Psychology, 19*, 21–39.

Hartig, T., Mang, M., & Evans, G. W. (1991). Restorative effects of natural environment experiences. *Environment and Behavior, 23*, 3–26.

Herzog, T. R., Black, A. M., Fountaine, K. A., & Knotts, D. J. (1997). Reflection and attentional recovery as distinctive benefits of restorative environments. *Journal of Environmental Psychology, 17*, 165–170.

James, W. (1950). *Principles of psychology.* New York: Dover. (Original work published 1890).

Kahn, P. H., Jr. (1999). *The human relationship with nature: Development and culture.* Cambridge, Mass.: MIT Press.

Kaplan, R., & Kaplan, S. (1989). *The experience of nature: A psychological perspective.* New York: Cambridge University Press.

Kaplan, S. (1995). The restorative benefits of nature: Toward an integrative framework. *Journal of Environmental Psychology, 15*, 169–182.

Kellert, S. (1997). *Kinship to mastery: Biophilia in human evolution and development.* Washington D.C.: Island Press.

Kellert, S., & Wilson, E. O. (Eds.) (1993). *The biophilia hypothesis.* Washington D.C.: Island Press.

Korpela, K. M., Hartig, T., Kaiser, F. G., & Fuhrer, U. (2001). Restorative experience and self-regulation in favorite places. *Environment and Behavior, 33,* 572–589.

Luhtanen, R., & Crocker, J. (1992). A collective self-esteem scale: Self-evaluation of one's social identity. *Personality and Social Psychology Bulletin, 18,* 302–318.

Markus, H. (1977). Self-schemata and processing information about the self. *Journal of Personality and Social Psychology, 35,* 63–78.

Maslow, A. (1970). *Motivation and personality* (2nd ed.). New York: Harper and Row.

Moore, R. C. (1990). *Childhood's domain.* Berkeley, Calif.: MIG Communications.

Morrissey, J., & Manning, R. (2000). Race, residence, and environmental concern: New Englanders and the White Mountain National Forest. *Human Ecology Review, 7,* 12–23.

Naess, A. (1989). *Ecology, community, and lifestyle* (D. Rothenberg, trans.). New York: Cambridge University Press.

Opotow, S. (1994). Predicting protection: Scope of justice and the natural world. *Journal of Social Issues, 50*(3), 49–63.

Opotow, S., & Clayton, S. (1998). What is justice for citizens' environmental groups? Poster presented at the meeting of the International Association for Applied Psychology.

Opotow, S., & Weiss, L. (2000). Denial and the process of moral exclusion in environmental conflict. *Journal of Social Issues, 56*(3), 475–490.

Oskamp, S. (2000). Psychological contributions to achieving an ecologically sustainable future for humanity. *Journal of Social Issues, 56*(3), 373–390.

Parsons, R., Tassinary, L., Ulrich, R., Hebl, M., & Grossman-Alexander, M. (1998). The view from the road: Implications for stress recovery and immunization. *Journal of Environmental Psychology, 18,* 113–140.

Ryan, R. M., & Deci, E. L. (2000). Self-determination theory and the facilitation of intrinsic motivation, social development, and well-being. *American Psychologist, 55,* 68–78.

Ritov, I., & Kahneman, D. (1997). How people value the environment. In M. H. Bazerman, D. M. Messick, A. E. Tenbrunsel, & K. A. Wade-Benzoni (Eds.), *Environment, ethics, and behavior* (pp. 33–51). San Francisco: New Lexington Press.

Rosenberg, M. (1981). The self-concept: Social product and social force. In M. Rosenberg & R. H. Turner (Eds.), *Social psychology: Sociological perspectives* (pp. 593–624). New York: Basic Books.

Scherl, L. M. (1989). Self in wilderness: Understanding the psychological benefits of individual-wilderness interaction through self-control. *Leisure Sciences*, *11*, 123–135.

Schultz, P. W. (2000). Empathizing with nature: The effects of perspective taking on concern for environmental issues. *Journal of Social Issues*, *56*(3), 391–406.

Schwartz, S. H. (1992). Universals in the content and structure of values: Theoretical advances and empirical tests in 20 countries. *Advances in Experimental Social Psychology*, *25*, 1–65.

Segal, P. S. (1999). The effects of nature-oriented and non-nature oriented guided imagery content on relaxation. Doctoral dissertation, Hofstra University, Hempstead, N.Y.

Sellers, R. M., Smith, M. A., Shelton, J. N., Rowley, S. A. J., & Chavous, T. M. (1998). Multidimensional model of racial identity: A reconceptualization of African American racial identity. *Personality and Social Psychology Review*, *2*, 18–39.

Tajfel, H. (1981). *Human groups and social categories: Studies in social psychology*. Cambridge: Cambridge University Press.

Thompson, L. L., & Gonzales, R. (1997). Environmental disputes: Competition for scarce resources and clashing of values. In M. H. Bazerman, D. M. Messick, A. E. Tenbrunsel, & K. A. Wade-Benzoni (Eds.), *Environment, ethics, and behavior* (pp. 75–104). San Francisco: New Lexington Press.

Thompson, S., & Barton, M. (1994). Ecocentric and anthropocentric attitudes toward the environment. *Journal of Environmental Psychology*, *14*, 149–157.

Triandis, H. C. (1995). *Individualism and collectivism*. Boulder, Col.: Westview Press.

Ulrich, R. S. (1984). View through a window may influence recovery from surgery. *Science*, *224*, 420–421.

Ulrich, R. S. (1993). Biophilia, biophobia, and natural landscapes. In S. R. Kellert & E. O. Wilson (Eds.), *The biophilia hypothesis*. Washington, D.C.: Island Press.

Weed, T. J. (1994). Central America's "peace parks" and regional conflict resolution. *International Environmental Affairs*, *6*, 175–190.

Wilson, E. O. (1984). *Biophilia*. Cambridge, Mass.: Harvard University Press.

Young, R. A., & Crandall, R. (1984). Wilderness use and self-actualization. *Journal of Leisure Research*, *16*, 149–160.

4

Human Identity in Relation to Wild Black Bears: A Natural-Social Ecology of Subjective Creatures

Gene Myers and Ann Russell

Writing in 1968, Erik Erikson suggested that the process of identity formation is an interplay between the individual and the communal, an ongoing dialectic throughout the life-span. In this psychosocial process, the individual "judges himself in light of what he perceives to be the way in which others judge him in comparison to themselves and to a typology [of others] significant to them" (1968, p. 22). This process is much like that described as "reflected appraisal" by other social psychologists (Cooley, 1902; Mead, 1962). Erikson also suggested the ideal direction of this development: "At its best it is a process of increasing differentiation and it becomes ever more inclusive as the individual grows aware of a widening circle of others significant to him," ultimately reaching the level of "mankind" (1968, p. 23). Achievement of identity, then, is a process of greater realization of who one is, and this process of comparison at the same time reveals one's relatedness to others.

The sweep and articulation of Erikson's theoretical vision still impresses us today. When we look at his vision now, however, we can ask if Erikson was wise to draw the furthest circle of identity at the human species line. Identity, denoting so strongly the scope of a person's sphere of concern, seems a promising concept for understanding peoples' investment in nonhuman nature as well. In this chapter we examine the possibility that the "communal" side of identity formation reaches to wild nature as well. The human-social and the natural-social communities of identity are distinguished chiefly by whether the "others" with whom we relate socially are human or nonhuman animals. We explore this distinction, and the processes that support it, by analyzing the reports of ten individuals' experiences with wild American black bears.

In the temperate latitudes, South America, and the Southeast Asian tropics, bears are the subjects of folk tales that imbue them with personality and influence on human life (Hallowell, 1926; Shepard and Sanders, 1985). For humans who know North American black bears well, they offer rich interactions. In the bear-people management career of one of the present authors (Russell), the individuality of bears has been notable. In captive contexts, according to Burghardt, black bears are capable of great behavioral complexity and adaptability to humans, and even of "mutually close, trusting, and playful relationship[s] between experimenter and bear" (1992, p. 377). Such experiences prompted the question—and admission, "Is not a *critical* anthropomorphism the next important step we [ethologists] are too wary to embrace openly, but covertly use?" (Burghardt, 1992, p. 380, emphasis in original). The black bear is as likely a candidate as any wild animal in North America for finding human identities that are vividly shaped by a relation to it.

If the human relationship to black bears can affect a person's identity, identity has relevance in bear-people management. The United States and Canada report healthy and in some cases increasing black bear populations. Conservation groups, however, are concerned about these numbers because of increasing human populations in bear habitat, poaching, and loss of forested habitat. Wildlife management agencies are increasingly dealing with human–bear conflicts. The number of "nuisance" bear reports has risen recently in several states, including Washington, California, New York, and New Jersey. Although concern for people's safety can outweigh tolerance for bears, most people are reluctant to see bears destroyed. Many want to understand bears better and desire to coexist with them. Whether we achieve coexistence may ultimately depend on our sense of identity in relation to them.

Background

In this chapter, the identities of people who have had frequent encounters with wild black bears provide insight into the existence and shaping forces of one sort of environmental identity: an identity anchored weakly in society, but strongly in a nonhuman animal presence. The ground for this anchoring lies in our social development.

Child Development

Evidence and theory from child development indicate that interspecies interaction may be a rich context for the development of self. The central act in the social genesis of self and identity is one of reflected appraisal, of seeing the self through the eyes of the other. The other's word or action provides an occasion for an act of self-reflection and realization of "who I am" in relation to the other. For this to happen, the other must be perceived as a social and subjective other, with a point of view on the self that is conveyed by something in its behavior.

The traditional view is that these requirements are only met between linguistic partners. Research in child development, however, suggests that this view is too restrictive. Building on empirical studies of infancy, Stern (1985) distinguished four layers of self–other relatedness, each of which is based on specific interactive abilities. By 3 months a "core" self jells as the infant perceives that self and other continuously possess agency, bodily coherence, and affect. Later, cues from caregivers establish an intersubjective domain where attention, affect, and intention can be shared (or fail to be shared), and thus crystallized as part of the self and other. Finally, another layer is added, based on verbal attributions of self and other, and linguistically constituted categories of behavior and identity. These domains of relatedness emerge sequentially in early development. All continue, coexist, and interact because each is based on invariant or unchanging dimensions of interaction that continue across the life-span. Thus what is possible for young children reflects lifelong developmental potentials.

In interactions with animals, preschoolers use all the capacities Stern identified (Myers, 1998). They relate to animals, but they also differentiate them from people within each domain. Our species' early social abilities are flexible in order to allow accommodation to other species' behaviors. Study of children's pretend or imitative play as "animals" shows that translation of one's own body into an animal's is an important way that relatedness becomes conscious, without depending on language. Similarly, and significantly for the question of reflected appraisal, behavior of the other is "read" for its response to the self.

The available mixed-species community thus provides the universe of social interactions in which the child finds comparisons and reflected appraisals of self and other. Given some richness of interspecies interaction, early human identity is situated, not just in a human-social world, but in a natural-social ecology of subjective creatures.

Identity and Wild Animals in Adulthood

Studies of adults' interactions with pets and service animals suggest that animals can be highly significant social others across the life-span in a range of societies (Podbersek, Paul, & Serpell, 2000). Of particular interest are studies of adults who work closely and sensitively with animals of a certain species or variety. In such cases, experience and knowledge build upon the basic interactiveness observed in children to result in performances with high degrees of meshing between partners' behaviors (e.g., Hart, 1994; Sanders, 1999). Such extensive interaction can hardly leave identity unaffected.

Little is known about naturally occurring adult–wild animal interactions, especially about their psychological dimensions. Fiedeldey (1994) found that seeing animals in the wilderness was of only moderate importance to hikers. DeMares' (2000) examination of human–wild cetacean interaction showed marked effects on the sense of self. Between the extremes of indifference and ecstasy represented in these two studies lie widely diverse experiences, but researchers have hardly begun to tap all the variations.

Other evidence of adult relations to wild animals comes from the less systematically psychological but more abundant sources in the writings of naturalists and in reports of anthropologists. In these cases extended familiarity with the given species may lead to identification, as revealed in the use of mental attributions to, and the deep concern for, the animals.

Cultural worldviews greatly elaborate Stern's (1985) verbal domain of self–other relatedness. For adults, worldviews must be considered in our question. The symbolic meanings these systems give to animals affect people's perceptions of animals, and in turn make the animals useful as identity symbols.

Hunter-gatherer groups present the most complex systems; in these societies elaborate worldviews centered on wild animals coevolved with intimate familiarity and dependence on them. Some of these systems are premised on a spiritual-moral continuity stretching across time. In such worlds, Ingold (1994) points out, animals are autonomous, but humans also depend upon their willingness to sacrifice themselves. Thus humans must earn their trust. "The animals . . . are supposed to act with the hunter in mind," and "have the power to withhold if any attempt is made to coerce what they are not, of their own volition, prepared to provide" (Ingold, 1994, pp. 14–15). Note that the human self receives appraisal from the animal other, in much the same process of judgment that Erikson indicated. Survival depends on an ability to tune in to animals' states of mind. Commenting on Bushman hunters' bodily felt sensation of being in the prey animal's body, Serpell holds that it implies a "profound level of identification with animals" (2000, p. 115). Such worldviews mobilize human capacities for a self–other relation with animals to a high degree.

Humphrey (1984) observed that the intensely social context of our recent evolution put a premium on anticipating others' behavior by a self–other comparison. If this propensity were to ignore species boundaries, the possible adaptive value is obvious. If the roots of self in relation to animals are as evolutionarily and developmentally robust as they seem, it is unlikely that identification with animals would be restricted to cultures with belief systems that facilitate such identities. Rather, where the requisite interactions with animals are rich, some type of self-in-relation may arise. Instead of preventing or distorting such identification, an anthropocentric and mechanistic-materialistic culture that lacks linguistic distinctions for animal subjectivity might merely ensure that it remains inarticulate.

If identity in relation to animals may be a lifelong developmental potential, does it occur in adults in contemporary American culture? Moreover, can it occur when the animal in question is wild, and thus neither affords the densely meshed interactions that domestic animals do, nor conforms to human expectations? What are the contributions of firsthand interaction with animals versus human influences such as social group and worldview? This chapter extends previous work in these critical directions. The question is whether, in direct experiences with the natural environment where the influences of social identity are relatively weak, human identity can authentically transcend the boundaries of our own species.

Study Methods

Subjects

Eight male subjects who had intimate familiarity with black bear species in the wild—familiarity built on different sorts of interactions—were sampled in four different categories of black bear-related expertise. Two other males with less contact and knowledge were also interviewed.

• Miles and Harold: These are brothers-in-law more than 80 years old who live in the river valley of the Cascade Mountain foothills in Washington State, where their fathers or grandfathers homesteaded. Especially in their earlier years, bears were simply part of their environment.

• John and Bob: These are hunters with broad interests in wildlife who watch bears more often than kill them. Both hunt bears with bow and arrow. The other subjects had hunted, but not extensively nor in this manner.

• Mitch and Clyde: These are wildlife biology bear specialists, masters of the scientific literature, but also possessors of massive amounts of firsthand experience. They are also educators.

• Rick and Rex: These are naturalists who enjoy studying animals firsthand and by tracking and reading animal "sign," a skill taught by a nature study school with which both have been involved.

• Sam and Charles: They are homeowners who reported nuisance bears near their homes to state wildlife authorities. While they are more knowledgeable about bears than many suburbanites, they represent a contrast here to the criterion of extensive familiarity with bears. In the past Sam hunted deer with a rifle.

All the subjects were residents of Washington State living in rural areas (the oldtimers, hunters, and Rex), suburbs (homeowners), or smaller cities (the biologists and Rick). Because of the difficulty of locating women in all of these categories, and because including them in only some categories would confound comparison across categories, only men were included in the sample.

Interviews

The interviews were conducted by phone or in person by one or both of the authors. They lasted 45 to 90 minutes and were taped and transcribed. We aimed to find whatever depth of connection our participants felt to bears, as they expressed it. The interviewees were first invited to tell a story of being in bear country. The story then provided points of departure for probing their identity in relation to bears (along the continuum of social influence) and identity formation processes such as reflected appraisal and relationships with bears. We also inquired about how bears affected the subjects' experience of the outdoors, their ideas about how people can understand bears, and their sense of self in community with nature.

Bears and Identity

The question of whether black bears enter into these subjects' identities can be addressed in two ways. First, how do the subjects' associations with bears form part of their social identities with other humans? Second, have their relations with bears affected their identities directly, with little mediation through the perceptions of other people, and if so, how? We view this second possibility as identity in a natural-social ecology of subjects.

Human-Social Identity

The natural environment clearly entered these people's sense of themselves as reflected in other humans' eyes. One hunter, Bob, described hunting as a tradition of three generations in his family. He never misses a deer and elk season: "I feel funny if I don't have an animal to eat." Bob's social relations are built around his wildlife pursuits. His wife hunts with him sometimes, and they moved to an area rich in wildlife. The other hunter, John, showed how identity can be bound up with positions on controversial issues that involve nature:

It scares me to death that a lot of people are trying to take hunting away. Like the trapping that was just banned in Washington. Because of that I'm going to end up leaving this state. I couldn't live somewhere I couldn't hunt or fish or trap.

Bears were also part of the biologists' social identities. Mitch described the almost stereotyped responses he receives from strangers when they learn his profession: "There's six bear jokes out there that I've heard perhaps a hundred times each." The implicit reflected appraisals in these examples reinforce identification with a social role—"hunter" or "bear expert." Clyde had assumed his role so thoroughly that in response to interview questions about his own perceptions, he often gave advice and information.

Although the preceding observations focus on bears, a strong socially mediated relation to the natural environment more generally was the backdrop in all cases. This is most clear among the rural octogenarians, Miles and Harold. Social identity and natural setting were virtually one for them. The immediate landscape was populated with memories and embodiments of social history: school sites, homes, graveyard, fields, fence lines. Their stories of bears used the same reference points as their family tales.

The relation to nature was pervasive in the social identity of the naturalist-trackers also. Here, however, the connection to nature was so strong that it created identity-management challenges. Here is Rex, the tracker:

> In actually seeing the animal itself, there's just this deep sense of an understanding. . . . it's on that feeling level, it's not [something I can] scientifically . . . just explain it to you. . . . Which is what I'm trying to develop as a really good solid scientist.

Confronted with others' misunderstanding of his incommunicable experiences of nature, Rex is now pursuing a science degree, in part to legitimate his views. A strong identification with nature in American society can leave a person feeling like a heretic (see chapter 14). Consider Rick's idea of what should be done about bears that intrude on human turf:

> My cynical response here? Sometimes I think, well, maybe it's not that there are too many bears but that there are too many humans. On the other hand . . .

Rick was wary of being perceived as pronature and antihuman, yet he also explicitly rejected anthropocentric values. Being strongly identified with the wild is not easy in a society that is increasingly distanced from nature. Both Rick and Rex have associated with a school of likeminded others; such a group may both support and enhance their identification.

Encounters with intruding bears had put Sam and Charles, the homeowners, in contact with agencies that manage people and wildlife in an alternately flexible and rule-bound fashion. In Charles's case, agency indifference to the bears increased his solicitous feeling for the animals, suggesting how a strong social context can mediate experience. Furthermore, since they did not have extensive firsthand familiarity with bears, the suburbanites' interpretations of bear behavior borrowed constructs from their social worlds, as we will see.

Thus there are several ways in which social forces can affect how identity incorporates the natural world, in this case black bears. As people with different pursuits, professions, pasts, passions, or perceptions, the subjects must negotiate these aspects of the self with an array of human actors.

Natural-Social Identity

As our introduction suggested, there are good reasons to suspect that nature enters human meaning and identity directly, as well as being mediated through the perceptions of other humans. This is particularly apparent through our extension of social connection to other animals. Experiences with nature, after all, are the raw material for the social attributions indicated above; for example, John, the hunter says:

Yeah, a lot of the guys think I use bacon grease for after-shave because I have a lot of encounters with bears.

How is identity experienced in relation to black bears themselves? Several variations emerged. One expression of identity is the person's awareness of a trait of the self that a relationship with the other has brought to light or clarified. One example was the realization that the human self is sometimes not the dominant creature:

Mitch (biologist): The value in having creatures such as grizzly bears and black bears in an area? Um, humbling.
Rick (tracker): I don't see bears very often, and I like to. It made it feel . . . probably in some ways I felt smaller.
Rex (tracker): [That night] I did not see or hear any of those bears but I knew that they were in the woods. And then I probably walked right past them again, and they left me alone. But it was humbling, a very humbling experience.

Wild bears add humility to modern human identity because of the unaccustomed subordinate role they can place humans in. This was remarked on by most subjects. On the other hand, the hunter, John, said his relationship with his prey revealed a dimension in his own nature:

I love the pursuit. I think it's something that's as deep rooted as the wolves are, or whenever they're chasing prey the same way.

John seems not to be expressing arrogance so much as a commonality with nonhuman nature. This dimension of his identity could not have been clarified without his special relation of hunter to hunted.

A relationship can enhance identity by contrast, even to the degree of placing the other in an outgroup. Identity crystallizes when one concludes that the self does *not* share some quality of the other, as in this humorous example from Harold, the oldtimer:

> I know friends of mine were out picking berries and she was talking to Warren her husband and he wasn't answering, he was just on the other side of the patch from her. So she walked around the other side of the patch and there was an old bear there feeding!

This punch line of a story on mistaken species identity plays on an assumption of the normal categorical separateness of bear and husband. In many details as well as in category, bears and humans contrast, but they also blur.

Another kind of example demonstrates how momentary interactions with the other, especially if repeated, can leave an indelible imprint, an experience that becomes part of who one is. Many people readily tell their bear stories upon hearing mention of bears, but our subjects told of experiences so profound they had shaped the person deeply. John could recall each of the seven bears he has killed: "I can think of each one that I've taken. I can still see most of them. It's hard to explain how I feel."

Experiences with bears in childhood or youth reverberated decades later. Even before we started the tape recorder, Miles pulled out a faded picture taken in 1930 when he was twelve—70 years earlier—showing him next to a huge black bear. The bear had been shot for killing eight or ten hogs, and Miles recalled the event nearly to the day (it was Easter or the day before). This bear was part of the mixed species community in the valley where Miles' emerging identity was formed. Biologist Mitch told how as a youth he accompanied a biologist studying bears at a town dump on moonlit nights:

> Just me and him. Bears scattering like flocks of seagulls or something, through the trees and across the dump. Never heard such large twigs and crunches and trees snapping underneath the weight of the very well-fed bears. Quite an incredible first experience.

Such experiences, when they occurred during development, or when they catalyzed a decision or attitude, deeply affected the eight expert subjects' identities. Bears and wildlife generally had come to hold great importance for these people.

In other cases, the relationship created an identification, or established a sense of similarity with the other on a concrete or more abstract level. Suburbanite and erstwhile rifle hunter Sam was disconcerted by one concrete resemblance:

Bears, you know, they look a lot like people when they are skinned out, it's really gruesome when they are skinned out, they look like people . . . really creepy.

For him, an uncanny resemblance to the human body challenged what had perhaps been a more comfortable separateness. However, for biologists such as Mitch, identification helped them do their job:

I've been known to walk around on all fours. Why was the bear here? I have photographs . . . of us hands down and noses in the dirt kind of thing, and really behaving in much the same way as a bear would . . . thinking like a bear. Make like a bear. Understand the bear from the inside out, definitely, I always think about it.

Mitch translated his own body and mind into the bear's to solve specific problems in the field. In retelling a story of a fake charge, tracker Rick subtly but vividly imitated the actions of the bear. Later, he invoked physical and mental parallels between humans and bears:

I don't like to say humans are the most intelligent animal, but I identify very well with human-type intelligence. A bear footprint looks like a human footprint only they're wider and they have claws on the end. There's a human-like quality that I identify with I guess. I can see myself in the bear, or maybe doing the things the bear would do, so there's also this human-like intelligence in bears, like curiosity.

Here Rick used an anatomical similarity to metaphorically evoke a psychological one. He also qualified what could have been a literal identity by noting his own biases. The result is to affirm likeness while also respecting unfathomed difference, giving a perspective on what it is to be human.

Finally, as we will explore in greater depth shortly, the respondents in varying ways expressed an ethical relation to bears. This was particularly true for the eight with high levels of bear familiarity, suggesting that conduct in relation to bears tapped into a deep sense of themselves as moral individuals.

Identification with bears is not surprising; indeed bears are among the more "anthropomorphic" of North American mammals in size, ability to stand on two feet, inquisitiveness, and dexterity. Moreover, many symbolic representations play on their human likeness. From teddy bears, to Yogi bear, to Russian dancing bears, and to more esoteric bears such as those of the bear cult of the Ainu, bears are peoplelike. Yet we would be in error if we interpreted our subjects' senses of identity in relation to bears as projections derived secondarily from the human-social dynamics of identity formation. To make our case that identity in relation to bears is generated through a relationship with the nonhuman species, we need to examine how well interactions with bears fit—and diverge from—the processes that generate self and identity within the human-social sphere.

Underlying Generative Social Processes

For bears to enter into the process of reflected appraisal that generates identity, and thereby to justify the expansion of identity formation from human-social to natural-social, three conditions are needed. The first is that they be perceived as subjective, as experiencing beings; the second is that they are felt to have a point of view on the self; and third, this is communicated by the bear's behavior.

Bears as Subjective Beings

Evidence that bears are perceived as subjective comes in the form of our subjects' ubiquitous attributions of intentionality (or mentality), intelligence, and individuality to bears.

Bears were described intentionally as perceiving and knowing. The subjects talked about how bears perceived such things as food sources, pathways, humans, and other species. They described bears as sensing by smell, touch, sound, and sight; performing intentional acts such as wanting, knowing, thinking, doing things for a reason, being puzzled, realizing, exploring, figuring things out; being "trap-wise"; needing "personal space"; and having qualities such as evasiveness, opportunism, playfulness, curiosity, and aggressiveness. These subjects used these terms in quite complex descriptions of bears, where they helped predict behavior. Consider the following description, from the biologist Mitch, of bears in a wildlife park who were introduced to human food in a simulated campsite:

It took them about three hours to *realize* that there was a source of food there. They kind of *explored hesitantly* at first, and then started to *realize* that there was more and more food available *to them.* (emphases added)

A year later, a tent was again introduced at the site, and the bears immediately came over to it. This example was thick with mental concepts and was offered to illustrate bear intelligence.

Indeed, intelligence was prominent in conceptions of the bear mind, as in the earlier excerpt from Rick. Bears' learning, adaptability, complexity of behavior, and related traits were mentioned across all the groups represented:

Harold (old timer): I think they are more intelligent than what . . . they have a mind there.
Rex (tracker): incredibly intelligent animals. In some studies, compared to dog pups, bear cubs are just way smarter.
Clyde (biologist): I found bears to be one of the most incredibly adaptable critters that I've ever worked with.
Charles (homeowner): I think they do think things through and reason. It's not all just instinctive reactions.

Hunter Bob deemphasized bear intelligence, stressing habit. But he also noted behavioral complexity: "About the time you think you've got it all figured out they do something different."

Mitch explicitly linked intelligence to another key mark of a social subject, individuality. As he put it, "just seeing the difference between individuals is, I think, another measure of intelligence." A species that learns extensively will exhibit behavioral variability across its members owing to the effect of divergent experiences. Later he commented, "You know, bears vary as individual humans do? They each have their own requirements as far as their psychology is concerned." Hunter John explained how individual differences in bear disposition add to the thrill of being around bears: "That's one of the rushes that I get out of being that close to such a big animal that, you know, there are individual animals out there that wouldn't mind eating you." Oldtimer Miles told about a singular bear who stopped traffic to beg for food by lying in the road. Perceiving bears as enduringly unique individuals deepens their basic subjectivity to create a personality.

The Human Self as Seen by Bears

Subjectivity is also tied to the second criterion of bears being seen as significant social others: seeing the other as having a point of view on one's self. Bob's talk reflected a habit of calculating what bears think. He used to leave home and, "this bear would come in and raid our garbage cans. He knew when we would leave. And he would never do it [if we were home]." Indeed, Bob viewed bears as calculating too, explaining how they restrict their movements when people, especially hunters, are around. Getting close enough to bears to shoot them with a bow is challenging, and he modified his own behavior to deceive them: "They can't identify you if you're crawling along or crouched over." In another passage, Bob described approaching a bear on a hillside: "He heard me so he knew what I was, but he had to get into a location where he could see me and I was trying to get into a location where I could see him." This depiction of an emotion-charged dance of mutual appraisal and intention reading typifies concern with another's perception of one's self, including here the human form.

It is no surprise that people think about what bears may think about them, for what bears think could have serious consequences. The eight expert interviewees said they felt a healthy respect for wild bears. As John (the hunter) said, "There's no doubt in my mind that if a black bear intended to eat you that it could get the job done." Both biologists and both trackers differentiated rare predatory or aggressive behavior from other types. Another special class of "problem bears," those who have become habituated to human presence and food, were regarded by all as untrustworthy. They are "one step closer to potentially becoming aggressive in their search for food in human areas," said biologist Mitch. All the subjects said they would fear bears in some circumstances.

Caution and even fear around bears are what we would expect based on popular portrayals of this animal. In contrast to the experts, Sam, the suburban homeowner who reported a problem bear, reflected this assessment. Although he was aware that wild bears are "leery of people because they're not domesticated," his interview was peppered with overwrought cautions about bear aggression, attack, and threat; about bears relishing people, harming children, being armed, angry, and dangerous. He felt an expert, upon seeing a bear, would behave this way:

Look around you, do you have rocks, do you have your gun on you, do you have a big stick, what do you have to protect yourself? Are you being real noisy, looking at him and all-"Heh! Get out of here!"

What is surprising is that fear was not a major element in the other subjects' actual reactions to wild bears. This was explained by a consistent theme in how they described bears' perceptions of humans: fear. Miles described how, while riding his bike at age 14, he "about ran into a bear . . . but he was frightened and I was too, I guess," and the bear ran into the woods. In story after story from his long life, Miles described bears that fled from humans. This crystallized the belief that "I'm sure they are looking for a way to get away from people." His brother-in-law Harold related similar episodes: "The bear took off, he was as scared of her as she was of him!" Bob described black bears as "so darn shy" they avoid people more than do other animals. Biologist Mitch said they are "very, very timid, and very anxious to get away from you." And Clyde said that even when radio tracking bears he seldom sees them; he can tell by the instrument when a bear detects him and starts moving away—generally well before he gets close. Hunter John described one bear he surprised in a tree: "I think it scared him worse than it did me, because he . . . he came right back down and took off running. He never stopped woofing." Naturalist-tracker Rick described a bear that approached his site at an alpine lake:

It was at least as surprised as I was. It wasn't sure what to do, it was kind of curious about the whole situation, and apprehensive about approaching closely.

We can use the subjects' term *respect* to sum up this response of the self to how the bear perceives the person. Respect combines awareness of the bears' potential to do harm with confidence born of experience that usually harm does not occur. Several subjects differentiated fear from respect, as Clyde the biologist did:

I've been around them enough that . . . there's not a sense of fear in me . . . [but] in many occasions, I've been reminded why we respect them so much, because they're wild animals, they are very very powerful, but yet they forgive our mistakes most of the time.

Respect implies correlative features of human identity, such as the humility noted earlier. That this feeling was held in common has significance for our larger questions because it contradicts cultural stereotypes of bears, because it cuts across the subjects' bear-related pursuits and across their worldviews, as we will see. Nor does this respect seem identical to (and thus simply derived from) any stance universally taken toward any discrete group of humans. It makes sense to see this commonality of respect as deriving from invariants of wild bear–human interactions.

Responding in Direct Interaction

The subjects' descriptions of bears' perceptions (of the self) show that move by move (in interaction), the animal's behavior was informative about its response to the human. For example, biologist Mitch told about meeting a grizzly mother and two cubs on a trail. Reading the mother's postures and actions, he interpreted her response to the situation as calm, in contrast to that of the cubs, who were running about "like dynamite." His description illustrates an on-the-spot synthesis of knowledge and reading of body language. It also suggests how different responses by individual bears generate different reflected appraisals and differentiated identity; the human had one meaning for the mother bear and another for the cubs. Just as we saw earlier how a person's own body can play a role in understanding and representing an animal, subtle reading of the actions of bears helped Bob, the hunter, succeed in interactions:

When I see an animal . . . I usually look for their body language. . . . So you can kind of tell if they're going to run, or if they're gonna hide, or if they're gonna be still so they think you won't notice them. . . . I look for their body language more than I think about their mental capacity.

These experienced observers of bears found the animal's body language readily intelligible.

Effective reading of behavior comes from informed intuition. The subjects familiar with bears drew on large stores of firsthand and shared knowledge. The suburban homeowners Sam and Charles, on the other hand, also read immediate reactions, but lacked critical knowledge. On the occasion Sam reported a bear, "The bear started pacing . . . that was weird. He was challenging me. He was like, 'wait a minute I'm not gonna run away this time'." This may be a reasonable, if coarse, assessment of problem bear behavior, but Sam's assumptions about how harm could result seemed to spiral out of control:

It doesn't matter if it's bears or bad guys. . . . These people [who get hurt by bears] are gonna be victims. And they're gonna get chased by bears and attacked by bears and their purse is going to be stolen and they're going to get assaulted.

Sam assimilated bear attacks to a human psychology of victimization. In contrast, when a bear pilfered his garbage, Charles read its behavior differently:

I decided to walk toward the bear to get a better look at it and then I realized I had my running shoes untied and I decided to tie my running shoes and that sort of broke the spell and the bear started sniffing around and I tied my shoes. . . . I didn't feel any fear, I felt curious and honored. . . . I wanted to go take it [some food], sort of honoring and recognizing this wild, this wild, I don't want to say animal, this wild being.

Although it is almost diametrically opposite to Sam's response in emotional valence, Charles's reading of the bear's behavior is similarly undifferentiated from human psychology. By their deficiency, these two subjects underscore the importance of knowledge and firsthand familiarity in a sensitive reading of behavior and responses of respect.

In a way, the same processes are working in the suburban homeowners as in the other subjects to shape identity in relation to bears, but something is clearly different. In the absence of intimate familiarity with the animals, social influences color the reflective appraisal process, precluding a more authentic identity in relation to the nonhuman.

Worldview and Identity: Ethics and Community

So far we have attempted to present a coherent perspective on how identity may be formed in a natural-social ecology of subjects, the "subjects" here being black bears and humans. The relations of the subjects who were familiar with bears show all the requisite processes for authentic identity formation, and the results, such as humility and respect, are not consistent with cultural stereotypes of bears as dangerous. We have argued that this commonality has roots in first-person interaction. If these points hold, then human identity can be authentically shaped by relation to at least this element of the nonhuman environment.

Identity, however, is inevitably tied to its human-social roots too. This is shown by how two intertwined aspects of our respondents' identities— ethics and membership in a natural community—were shaped by cultural worldviews. Worldviews, sometimes in the form of religions, are culturally transmitted comprehensive sets of assumptions about the universe, causality, humanity, ethics, nature, and supernature. They cannot be entirely falsified by experience, and so exert their influence—albeit an incomplete influence, we will argue—upon it.

The Religious Worldview

The 80-year-old subjects described earlier brought with them a traditional Christian cosmology when they went into the woods. It supported their equanimity in killing harmful bears, and helped define the separateness of a settled human community and a woods-animal community. God's design helped them interpret natural phenomena, but as one of them indicated, the idea of God was also enriched by their experience of animals: "It's interesting how God has instilled in the minds of these different animals all the intelligence that they need." Harold described why he likes to hunt: "Well, I think just getting out there; walking along I catch myself not really looking for the animals there, just looking to see what God has created and enjoying it. And yes, the meat we got was used."

As the last remark shows, religion grounds a particular conception of a natural community as separate and unequal. Miles summed this up: "Still I'd like to see them have their part of the woods and stay there." However, this view is not lacking in ethical sentiment. Miles, Harold, and John the hunter all grounded their ethical judgments in Christian religion: "The Lord has given us these animals and as He told Peter to kill and eat so, I think if it is done in a humane way, I don't think it is too bad" (Miles). John shared this perspective, for example in commenting on a beaver he had trapped: "It's a beautiful animal, soft pelt, but my beliefs are that they were put on the earth to be used."

It must be emphasized that this Christian theme was not used to justify ruthless exploitation. John realized other people might think a hunter cannot respect animals, "but I know that that's not true, and I know that I respect animals." For him, this meant leaving a renewable resource and not overhunting, as well as something harder to define about the animal itself. He told how "a lot of times I'll say a little thanks" upon finding his fallen prey. Respect is carried forward here as awe, and as a humane and utilitarian conservation ethic—albeit one that gives humans management and survival prerogatives.

The Scientific Worldview

The biologists were the most circumspect about any subjective dimension of community. They articulated a trained uneasiness with the slippery slope of anthropomorphism; it can become the "fuzzy-cuddly approach," as Mitch put it. However, explicit in what we have seen is that bear subjectivity must be acknowledged, if only in practice. Mitch avowed, "you can't help but try and interpret [bear] behavior in ways that you might try to interpret human behavior, using the same tools in your mind. That's all we've got." Both biologists used mental attributes when it helped make a critical educational point. When asked for examples of community, however, Clyde remained agnostic, and Mitch gave examples that required little mental inference. It was clear they had a vast understanding of the "community," but thought of it in objective terms. This was completely compatible with their strong ethic of human—animal coexistence. Respect here generates a demanding obligation to be diligent in meeting the bears' needs. This sense of coexistence means material, not psychic, compatibility.

An element of the modern biological worldview—evolution— characterized not only the biologists' perspectives but also that of others. Evolution blended with the religious dimension for John, who, as we saw earlier, felt ancient echoes in his pursuit of prey. He added "I get a lot of satisfaction out of taking a deer with a bow. More like a religious experience, I guess." His concept of human nature is engaged here: "I think [the desire to hunt is] probably stronger in some people than others. . . . In the right situations . . . you get in a survival situation and a lot of people change their views." The evolutionary past was thus an element in his sense of the rightness of hunting. In a broader sense, evolution also characterized the worldview of the trackers, who portrayed a profoundly different variation of natural community.

The Ecocentric Worldview

The trackers, Rex and to a lesser extent Rick, talked about bears in a very different way than any of the others. Rex not only directly attributed intentional states to them, he also spoke as if he habitually assumed that bears are knowing, active entities in the woods. He described how they "schooled me, no matter how prepared I was," for example. This repeated term, "schooled," implied that bears deliberately administered lessons to him. His descriptions of bear behavior were less concrete than those of the others, and more suffused with a sense of psychic communication. Rex described meeting several bears at night on a trail:

I absolutely had my very calm state of mind and passive awareness, soaking up the fact that I was in the woods with these beautiful creatures. . . . And I said "Hey, hey bear I'm out here and I'm just going to pick up some tracks. I plaster casted them earlier, maybe they're yours, this is how I learn." And I was just very directly talking to the bear and then I kept moving. And the bear kept moving in the opposite direction. . . . And there was a definite little quiet turning point where I realized I was going to keep going and I was going to be confident with it. . . . Just basically talking to them in a confident, friendly voice, in a nice stream. And the bears could sense my state of mind, I'm sure of that.

The assumption here is not simply that bears respond to behavior, but that they respond to attitude. For Rex an emotional epistemology works in relations to animals: "It's on that feeling level that that connection with an animal forms."

Not surprisingly, Rex and Rick gave the most elaborate and engaged descriptions of intersubjective ecology. Rick described a vocabulary of bird songs, and his aspiration was to understand them well enough to decode the complex interactions they were presumed to sometimes represent. For him, the effect of this awareness was to ' [make] me feel connected" to a natural community. This is

a very big feeling and a very small feeling, all at the same time, because there's a feeling like you don't end at the tips of your fingers or the top of your head. Because what's going on has to do with what you're doing, and with other things that are going on, so you're kind of a part of this whole thing, and very small because there's so much more out there.

A very strong ethic of egalitarian coexistence corresponded with this attitude, with respect taking on its most personal and mutual meaning for Rex:

[There are bears] in British Columbia that really have developed a taste for human flesh. I would say no. Humans aren't put here for bears to use; well, they're not put here for us to use either, but sometimes we eat each other.

Rex reported killing one bear, but described his reverently careful use of it. Participation in the ecosystem is complete.

This participation taps a particular worldview. The school of nature study with which both Rex and Rick have been associated draws inspiration from ecological science, as well as from traditional hunter-gatherer cultures. Although it lacks traditionalist social strictures and rituals, their school teaches that there are other ways to know about animals and nature than the scientific. Rex gave an evolutionary rationale for the ability to understand the psychological community of nature:

Through most of the time period of our ancestors as humans in the ecosystem, in the evolutionary ecosystem, [they survived] because of those skills and that kind of awareness. . . . It's hard wired in us.

Rex and Rick's views may have been shaped by this specific school's influence, but the ideas involved are also available in American environmental thought.

In attempting to understand this last profoundly divergent sense of self-in-community, it may be wise to recall cautions against uncritical anthropomorphism, including how it can put a stop to inquiry by providing plausible but untestable answers to questions about animal behavior. However, all worldviews are to some extent untestable, and there may be a return from adopting permissive assumptions. They may lead to new insights about other species because they encourage the fullest use of our social abilities rather than curtailing them by a too parsimonious skepticism. When coupled with humility and knowledge, such a worldview also offers enriching experiences of identity and new possibilities for coexistence.

Common Themes

Ethics combined with community generated differing conceptions of coexistence with bears, but interactional respect and common community worked within each concept. All eight of the expert respondents in our study expressed a strong respect for black bears. As the word implies, morality plays a role here, and it in turn was shaped by differing worldview commitments. Still, we suggest, the basic concept of respect is not constituted by worldviews, but rather works differently through and beyond each perspective.

All of the expert subjects felt some sense of common community with bears and nature. Both hunters, one biologist, and both trackers responded affirmatively when we asked if they thought of nature as a community of animals that were potentially communicating. As in the case of hunter-gatherer cultures, the experience of nature as a community of subjectivities is a possibility, but it appears to be mediated by differing worldviews. Again, however, we suggest that this is built upon interactional experience. For those eight subjects with expertise, bears or their possible presence heightened sensation. Several subjects reported that they attend to other animals' actions, particularly the alarm calls of birds, in the effort to find bears and to avoid letting bears detect them. Bear country is pervaded by bears as subjective presences, a perception that is not changed by any of the worldviews.

Conclusion

The human influences shaping identity in relation to nature—and to black bears specifically—are powerful. Traditional social science analyses of the impact of nature on identity might end here, with group identities, work roles, and cultural assumptions. However, we have asked the further question: Is nature directly formative of identity, constituting the self in a "natural-social" realm that interacts with but is not wholly molded by human-social factors?

With two exceptions, all our subjects had extensive experience with wild black bears, enabling us to query their identity in relation to bears. The subjects' identities were affected in several ways, for example, a clarified sense of what it means to be human. This natural-social identity springs centrally from social psychological processes such as reflected appraisal and direct interaction that give rise to a sense of self-in-relation. Such ingredients of identity, annealed with a worldview, allow the self to be situated in a morally colored natural community.

Woven into our investigation are larger questions. Based on child development research, a sense of self in relation to animals can be hypothesized to be a lifelong human potential whose main prerequisite is intimate familiarity with another species. Our findings support this potential, and show that it extends to relation to a wild animal. Not only was identity shaped in relation to bears, as we summarized earlier, but more particularly, interaction with the actual animal appeared to be generative of this sense. Specifically, in the subjects' words, respect characterized their regard for bears. Respect mixes awareness of the bears' power, confidence in one's own safety, and a norm of coexistence. The correlative feature of human identity was some variation of humility. We argued that this cannot be derived from cultural stereotypes, which show bears as dangerous. That these features held (with variations) across worldviews suggests a strong experiential basis. That they were common across all but one category of subjects suggests a derivation from invariants in wild black bear–human encounters.

In our analysis, these invariants come from the processing of species-typical black bear behaviors by humans' flexible capacity to read animal behavior for self-relevant cues. However, this process must build on extended familiarity or it may be distorted by interpretive frames generated in a human-social sphere, as shown by the deficiencies of the subordinates' readings of bear behavior. A similar flexibility of the reflected appraisal process is demonstrated by young children. The ability to read bodily activity for basic subjectivity and cues to quasi-shared subjectivity, plus interpretive skills that ultimately stem from language use are all deployed and—with knowledge and experience—appropriately accommodated in forming a sense of self in relation to other species. Thus, the same abilities that generate identity in the social realm of other humans extend to the natural-social realm as well.

The different senses of community with nature, and their associated kinds of human identity, would not be possible if we were not human-social selves who participate in cultural worldviews. Our subjects depended on untestable assumptions to construct community-level connectedness out of the raw materials of more immediate natural-social relatedness. Yet this influence of worldview did not override the commonality they called "respect." If we let respect carry with it the rich relational meanings given to it by our subjects, we can see how it organized their experiences differently in their different world views. Given rich experience, our basic social abilities can reach across species boundaries in interaction, affecting identity in ways that precede and surpass the influence of a particular culture. In its broadest implications, our study expands the circle Erikson (1968) placed around identity by drawing the human experience of identity in connection to intimately known wild animals and nature into the scope of our social self-in-relation.

References

Burghardt, G. (1992). Human-bear bonding in research on black bear behavior. In H. Davis & D. Balfour (Eds.), *The inevitable bond* (pp. 365–382). Cambridge: Cambridge University Press.

Cooley, C. H. (1902). Human nature and the social order. New York: Scribner.

DeMares, R. (2000). Human peak experience triggered by encounters with cetaceans. *Anthrozoös 13*(2), 89–103.

Erikson, E. (1968). *Identity, youth and crisis.* New York: Norton.

Fiedeldey, A. (1994). Wild animals in a wilderness setting: An ecosystemic experience? *Anthrozoös 7*(2), 113–123.

Hallowell, A. I. (1926). Bear ceremonialism in the Northern Hemisphere. *American Anthropologist* (n.s.) *28*, 1–175.

Hart, L. (1994). The Asian elephant-driver partnership: The drivers' perspective. *Applied Animal Behaviour Science 40*, 297–312.

Humphrey, N. (1984). *Consciousness regained.* Oxford: Oxford University Press.

Ingold, T. (1994). From trust to domination: An alternative history of human-animal relations. In A. Manning & J. Serpell (Eds.), *Animals and human society: Changing perspectives* (pp. 1–22). New York: Routledge.

Mead, G. H. (1962). *Mind, self and society from the standpoint of a social behaviorist.* Chicago: University of Chicago Press. (Original work published 1934)

Myers, G. (1998). *Children and animals: Social development and our connections to other species.* Boulder, Col: Westview Press.

Podbersek, A., Paul, E., & Serpell, J. (Eds.). (2000). *Companion animals and us.* Cambridge: Cambridge University Press.

Sanders, C. (1999). *Understanding dogs.* Philadelphia: Temple University Press.

Serpell, J. (2000). Creatures of the unconscious: Companion animals as mediators. In A. Podbersek, E. Paul, & J. Serpell (Eds.), *Companion animals and us* (pp. 108–117). Cambridge: Cambridge University Press.

Shepard, P., & Sanders, B. (1985). *The sacred paw.* New York: Viking.

Stern, D. (1985). *The interpersonal world of the infant.* New York: Basic Books.

5

Moralizing Trees: Anthropomorphism and Identity in Children's Relationships to Nature

Ulrich Gebhard, Patricia Nevers, and Elfriede Billmann-Mahecha

The identity of an individual human being seems to be composed of both variable and invariable components. If the same person is observed in different contexts or at different points in time, certain characteristics seem to remain constant and reflect a kind of coherence and continuity, while others tend to vary, at least partly as a result of exposure to various kinds of experience. It is this more variable component of identity with which we have been concerned in our research. We suggest that at least part of the process by which the identity of individual human beings is formed involves making comparisons, assessing sameness and difference, distinguishing oneself from some things and people, and identifying with others. This very basic cognitive operation of comparison is often taken for granted in discussions on the psychology of identity, an omission we wish to correct in this chapter.

According to most psychological theories, identity is rooted in our experiences with ourselves and with other people. The significance of other human beings for the development of identity was discussed in depth by Mead (1934) and Erikson (1950); more recently Giddens (1991) has described the process of reflexive and narrative construction of concepts of self. However, the significance of our nonhuman environment has been given very little attention so far. Plants, animals, wind, and water play at best an insignificant part in most theories of identity formation. In the object relationship theory of traditional psychoanalysis, for example, the relevant objects with which the child is thought to interact and to represent in its mind are always humans. As mentioned in the introduction to this book, the psychoanalytical approach of Searles (1960) represents a notable exception. Searles investigated the significance of the nonhuman world for both normal development and that of schizophrenic patients. His thoughts are based on the idea of a fundamental kinship between humans and their nonhuman surroundings which constitutes the context within which mental and emotional development and identity formation occur. His main thesis is as follows:

> The nonhuman environment, far from being of little or no account to human personality development, constitutes one of the most basically important ingredients of human psychological existence. It is my conviction that there is within the human individual a sense, whether at a conscious or unconscious level, of relatedness to his nonhuman environment, that this relatedness is one of the transcendentally important facts of human living. (Searles, 1960, p. 5)

This chapter discusses a phenomenon closely associated with Searles's theory, the phenomenon of anthropomorphism. Through anthropomorphic interpretation an external object such as a tree or squirrel is perceived as being similar to oneself and humanlike in certain respects. The quality of being alive and the ability to feel pain are aspects frequently referred to in ascertaining such sameness. When both are placed in the same semantic category, the observer's knowledge about him- or herself can be employed to better understand the nonhuman object, and features of the plant, animal, or ecosystem can be drawn on to better understand oneself. Thus the identity of both the object and the observer take shape. Furthermore, perceiving an object as humanlike may evoke feelings of empathy for the object that permit it to be regarded as something worthy of moral consideration. Our results indicate that anthropomorphic reasoning is frequently used by children up to the age of about 12 to justify protecting trees and other nonhuman objects from harm.

Because of these effects, anthropomorphic reasoning might appear quite attractive to those interested in promoting environmentally favorable attitudes and behaviour. However, our results indicate that caution is in order when considering ways to utilize this kind of thinking for educational purposes. Three points are particularly relevant in this connection. First of all, not all objects lend themselves to spontaneous anthropomorphic interpretation. Ecosystems, for example, were not readily anthropomorphized by children in our sample. Second, explicit anthropomorphic reasoning wanes quite distinctly as children grow older, for reasons about which we can only speculate. And finally, since naive anthropomorphic interpretation is not always in the best interests of nature, we will argue that the goal of education should be a kind of "enlightened" anthropomorphism that also takes the "otherness" of nonhuman nature into account.

Empirical Approach

In Germany a great deal of research has been conducted on the factors that influence environmentally relevant behavior, one of which is attitudes (see de Haan and Kuckartz, 1998 and Lehmann, 1999 for overviews). Although children's attitudes toward nature have also been examined in several studies (summarized in the sources given), the tendency is to focus on perceived deficiencies. In order to gain a somewhat different perspective on children's thinking, we have turned to Philosophy for Children, an approach developed by the Institute for the Advancement of Philosophy for Children at Montclair State University in New Jersey and further refined for research purposes by E. Billmann-Mahecha. Group discussions in which children talk about problems among one another are a crucial element of this approach.

While group discussions in contemporary qualitative social research have been carried out mostly with adults and adolescents, advocates of Philosophy for Children have been discussing philosophical problems with groups of children for many years using a modified version of the Socratic method (see Lipman and Sharp, 1978). The main goal of this method is to encourage children to reflect independently on a subject and express their views in response to questions. Experience with discussions of this kind involving philosophical subject matter (for example, Matthews, 1984) and preliminary results of attempts to use this form of discussion as an instrument for qualitative research (Nevers, Gebhard, & Billmann, 1997) encouraged us to pursue the method further. In short, it involves presenting a brief stimulus, in our case a story, followed by a discussion led in a nondirective manner by an adult facilitator. Unless otherwise indicated, most of the discussions were conducted in German and translated into English by Nevers.

Compared with an individual interview, a group discussion has the advantage that the children relate more strongly to one another rather than to the adult investigator and are thus, we hope, less oriented toward the investigator's expectations. They discuss the topic within the concrete scope of their own experience and in their own terminology so that the texts that emerge are more closely attuned to what children consider relevant. Of course the children also influence one another, but this can provide insights into social processes of negotiation, that is, into the communicative development of opinions, attitudes, and values.

The group discussions were recorded on tape, transcribed, and analyzed sentence for sentence according to the procedures of grounded theory proposed by Strauss and Corbin (1996). These include (1) open coding, (2) summarizing prominent themes, (3) reducing the empirical material further by means of synopses of the thematic summaries, (4) analyzing selected passages in depth, and (5) developing theoretically relevant categories.

We have examined discussions carried out with children and adolescents in three age groups, 6–8, 10–12, and 14–16. The choice of participants was not controlled and is most likely skewed toward city dwellers. To initiate a discussion, a story was read that involved a conflict of interest between children or adolescents on the one hand and a nonhuman natural entity on the other. Three different kinds of objects were selected: individual plants, individual animals, and complex ecosystems or landscapes. Each story centered around a conflict that might occur in the everyday experience of children. For example, one figure in a story is in favor of chopping down a woods in order to extend the grounds of a school, while another is opposed to the idea. Although the conflict in the story is not a "real" one, it is not strongly hypothetical either. Hypothetical dilemmas such as Kohlberg's famous story about a man, Heinz, who contemplates whether to steal medication for his sick wife, involve constellations that rarely occur in real life and least of all in the lives of children. They are used in philosophy and Kohlbergian research in order to focus the discussion primarily on purely cognitive aspects of the situation. Since we are interested in tapping the emotional aspects of children's relationships to nature as well, we have chosen a type of dilemma that children might indeed encounter in real life.

We now have an extensive archive of transcribed discussions, of which fifty-seven have been analyzed in greater depth, at least five for each of the three major objects and age groups. Since each group consisted of approximately 5 children, the total sample corresponds to about 285 children.

One of the stories with which we have been working is the following:

The Tree House

Peter and Sarah are planning to build a tree house. It's supposed to be a particularly fine one, in which several children can sleep and eat and read comics. Each child is to have his or her own corner in the tree house for storing candy and other private things. The tree they've chosen for their building project is a lovely old willow with big, roomy branches on an empty lot not far away. It's really the only tree around that's suitable for their purposes. The others in the surroundings are in private yards or parks. Besides, they're all too small. Flushed with excitement, Sarah describes her plan:

"First we have to blaze a proper trail to the tree by chopping away all those brambles. Then we'll cover it with flat stones. Of course, we have to clear the space beneath the tree as well. Otherwise we can't find things if they fall out of the tree house. We'll fasten a ladder to the trunk and saw off those two branches that are in the way so that we won't have any trouble climbing in and out. Then we'll nail boards between the branches to serve as railings and put in some kind of floors here and there. We might have to remove some more branches if we need more light in some places."

While listening to his friend Sarah, Peter's expression becomes more and more serious. He's known the empty lot and the willow all his life and is familiar with every inch of them since he often goes there just to be by himself. He knows exactly where different birds nest in the tree and surrounding brambles and can name all the plants surrounding the tree. He finds the willow particularly beautiful just the way it is and replies to Sarah:

"I've changed my mind. I don't want a tree house after all, and I don't want other kids to build one either. I don't want to change the tree in any way at all. I think we should just leave nature in peace."

Sarah is annoyed. She can't understand Peter's lack of enthusiasm and replies in a huff:

"I don't agree at all. First of all, we're not talking about nature but about a messy old abandoned lot. I've never seen the owner, and he certainly doesn't care what we do to the tree. The tree itself can't care one way or the other. Besides, children have a right to play, and that's the most important thing."

What do you think?

In assessing the responses of children to this story, it is important to note the following characteristics:

• In the story Sarah's position is anthropocentric and resource oriented since she would like to use the tree for certain purposes. Peter, on the other hand, considers the tree to be valuable in and of itself and advocates nonintervention. His position is nature oriented and would thus be termed physiocentric in German literature and biocentric or ecocentric in North American literature. The children's conflict also reflects the perennial controversy between conservationism and preservationism.

• The conflict involves an individual organism as opposed to an ecosystem. This aspect also relates to an ongoing debate in environmental ethics. Individual organisms have been considered to be irrelevant or only relevant as part of a greater system (see Callicott, 1980 and ensuing debate documented in Hargrove, 1992). This probably reflects a general tension in moral discourse between what has been termed "micromorality" as opposed to "macromorality" or personal versus transpersonal spheres of moral judgment (see later discussion).

• The story involves a conflict of interest in an everyday setting. It thus resembles the stories employed by Eisenberg and her co-workers in their research on prosocial behavior (see, for example, Eisenberg, Shell, Pasternack, Lennon, Beller, & Mathy, 1987). However, different types of interests are involved. The "interests" of the tree addressed in this story are quite basic ones, that is, surviving and remaining whole and undamaged. Those of the children are also quite basic—unrestrained freedom to play—but are not quite as essential as those of the tree since they could be satisfied in another manner. Asymmetrical conflicts of this kind are common in industrial societies, while conflicts involving essential interests of both parties are probably more prevalent in countries that are not as industrialized.

• The conflict is located in what Eckensberger (1993) might call a personal sphere of moral judgment as opposed to a transpersonal one involving society in general. It concerns what Rest, Narvaez, Bebeau, and Thoma (1999) would probably call an example of "micromorality" involving relationships on a one-to-one basis and partisan caring as opposed to "macromorality," which is more concerned with structures that enhance cooperation at a societal level and which requires impartiality and acting on principle. While macromorality is the focus of Kohlbergian studies, research on prosocial behavior is more concerned with micromorality. Thus in this respect as well our research more closely resembles that on prosocial behavior.

• The situation centers around negative duties rather than positive ones. The children discuss whether it is justifiable to damage the tree by building a tree house, not whether they should water it. According to Kahn (1999) a conflict involving negative duties as opposed to one involving positive duries is generally more readily conceived as being morally binding and obligatory.

The focus on an individual relationship rather than a societal problem, the asymmetry of the relationship involved, and the prospect of damage to the tree probably tend to evoke a view of the tree as being "needy" and stimulate a protective and prosocial response. Along the same line of thinking, studies by Susan Opotow (1987, 1994) indicate that in some cases perception of the "neediness" of a natural object may be more inclined to generate concern for its protection than perception of the object's potential usefulness. These observations may in part explain the intensely empathetic responses of younger children documented here.

Anthropomorphic Patterns of Interpretation

Moralizing Trees Through Anthropomorphic Interpretation

One of the most striking results of the group discussions we have analyzed so far, which substantiates previous studies by Gebhard (documented in detail in Gebhard, 2001), is the widespread use of anthropomorphic reasoning in dealing with conflicts with nonhuman objects. Not only animals such as dogs, squirrels, and hornets, but trees are also described and interpreted as if they were human. Familiar physical qualities of humans such as blood and hair are projected upon nonhuman objects by metaphorical analogy. The leaves of the tree are construed to be like human hair; the sap is compared with blood. Thus with reference to cutting off the branches of a tree, Billy (boy, 10 years old) says: "Because [if] I like saw people chopping off, you know, like, like I'd feel insane really like someone's sticking something into you, like or pulling skin off" (quotation originally in English). Nan (girl, 10 years old) rejoins: "Yes, sometimes I feel that feeling when like someone gets hurt, I just see it and then this feeling runs down my body, you know like, ahm, feeling that I kind of feel that too, how it hurts and everything" (quotation originally in English).

The children are interpreting the tree in terms of their own bodily experience, of which suffering and pain are very basic elements. Their anthropomorphic interpretation seems to be closely associated with feelings of empathy for the tree and appears to permit them to assume its perspective, at least approximately. Both of these qualities are known to be important aspects of moral development. The children's perception of potential pain and suffering leads them to reject their personal interests and advocate nonintervention, a decision reminiscent of a pathocentric position in environmental ethics. We propose that the children conceive of the tree as a moral object and that anthropomorphizing nature allows it to be moralized.

While analogies between humans and trees are particularly conspicuous with respect to anatomical and functional properties such as organs or the ability to feel pain, mental and emotional qualities such as consciousness, feelings of happiness or despair, and intentionality may also be ascribed to nonhuman objects by younger children (aged 6–10). Thus flowers droop when they feel bad and trees are sad when their skin is torn away. And: "A tree is like a blind person who's tied up and can't hear" (boy, 10 years old). "Can't hear or see. Yeah, the only thing a tree can do is feel. That's just the thing, feelings. A tree feels pain" (girl, 10 years old).

In the course of identifying with trees by means of analogy, the act of killing plants is compared to killing humans. This is demonstrated by the following particularly striking comment by a 10-year-old girl:

We can defend ourselves and run away, and the trees stand there helpless and can only wave their branches in the wind. But they can't stick up for themselves and swat you with their branches. The only way they might get back at you is if someone catches up with you under a tree who doesn't like you and has been chasing you. But trees can't defend themselves. We don't hang around either, waiting happily and saying, "Oh great! Here comes a murderer and wants to murder me!" Trees can see that coming and trees can be scared.

An important element of the arguments presented to justify protecting trees is the concept of life: "Actually, trees are alive too, and they have a right to stand there, and the bushes, they have the right to grow too" (girl, 10 years old). The rights of these organisms are thus associated with a very basic property common to humans and plants, namely, the condition of being alive. In fact, the terms *alive* or *living* are frequently coupled with moral arguments in discussions about trees and other organisms as well. Children argue that "nature wants to live," and "after all, trees want to live too." These arguments resemble those of the biocentric philosophers Albert Schweitzer (1995) and Paul Taylor (1986), although the children do not elaborate their position in greater detail. However, since the children apparently regard the condition of being alive as a decisive criterion for assigning intrinsic value, it seems legitimate to speak of a biocentric perspective. Kahn (1999) has made similar observations, but uses the term *biocentric* in a broader and more comprehensive manner.

It appears then that younger children are quite readily able to apply a moral rule commonly used for dealing with relationships between humans to a nonhuman object, a rule codified in the Human Rights Convention. It not only demands that life be protected (as in Article 3), it also defends the need for shelter (as in Article 25). Thus children argue against cutting down a forest by pointing out that the homes of animals may be destroyed. Kahn (1999) has termed this kind of reasoning, in which value or justice correspondences between humans and other natural entities are recognized, "isomorphic reasoning." It requires acknowledging the identity of a natural entity and recognizing a kind of symmetrical relationship between this object and the observer. From the standpoint of moral psychology, reference to moral rules in this age group and their application to plants and animals reveals an argumentative position that is more mature than might be expected according to Kohlberg's theory of moral development. Divergence from Kohlbergian theory is also indicated by the observation that although children may occasionally refer to adults as authorities, arguments in which reference is made to possible sanctions or fear of punishment are rare. Perhaps this corresponds to what Eckensberger (1993) might call autonomous moral judgment in a personal sphere.

While primary school children are quite clearly prepared to attribute mental and emotional qualities to trees, this more extreme form of anthropomorphic projection diminishes with increasing age. An organism's capacity for having feelings and being able to think seems to become more closely associated with being able to speak, as the following quotations from three discussions show.

"If a tree could speak, I'd ask it whether or not a tree house hurts" (girl, 10 years old).

"If a tree could think, then it'd really be alive" (boy, 10 years old).

"If a tree could speak, I'd excuse myself for not knowing whether or not is alive" (boy, 10 years old).

The phenomenon of identification with trees seems to give way to a more detached perspective as children grow older. Thus they may argue that a tree "Isn't really alive the same way as a person is" (girl, 10 years old). At any rate, among children aged 12 to 13 on up we found no explicit anthropomorphic analogies in discussions about trees. It follows that children 12 years and older express less empathy with trees and tend to question the ability of trees to feel anything at all. They also exhibit greater interest in building a tree house, and seek compromises that would allow them to do this without harming the tree or the animals that live in it. Here ambivalence is clearly evident. On the one hand, doubts about anthropomorphic interpretations stemming from new cognitive insights are expressed and result in more assertiveness in defending personal interests. On the other hand, anthropomorphic interpretations still seem to be active, as evidenced by the children's sincere efforts to find a compromise.

Among even older subjects (14–16) the idea of sentience among trees is no longer evident, or at least it is no longer explicitly expressed (although this does not necessarily mean that anthropomorphic interpretations cease to exist). Moreover, the 14–16-year-olds we interviewed were clearly less inclined to make any sacrifices for the sake of a tree. Instead, greater significance was attached to personal interests, such as mobility and having fun, and social interests, such as getting together with friends. Furthermore, societal aspects such as earning money and unemployment are mentioned more frequently than in discussions with younger children. Arguments in which the perspective of another person is considered are also expressed more often. Thus a tree or goldfish may no longer be considered valuable in and of itself, but it is respected if a friend is attached to it for some reason or other. These observations suggest that society plays a more prominent role in mediating adolescents' relationships to nature. Furthermore, it appears that rules that normally apply to humans, such as those of the Human Rights Convention, are no longer regarded as equally valid for nonhuman objects. In dealing with nature, other rules are brought into play, for example, animal protection or nature conservation laws.

We can conceive of several reasons for the decrease in explicit anthropomorphism observed in adolescence. One probably has to do with the construction of a more elaborate and hierarchical concept of "life" in the course of education and development. For adolescents, human life is different from animal life and both are superior to plant life, a view that is widespread in the general public. A second reason for less explicit anthropomorphism among adolescents may be that in this phase of their lives they become more concerned with their place in society and their social identity than younger children.

Views of Environmental Destruction in Discussions about Trees

In many of our discussions about trees, destruction of the environment is a central theme and views are expressed that reveal existential fears. The following quotations are from a single discussion:

When all the trees are dead, afterwards, . . . then nothing else can survive. (girl, 10 years old)

Without trees we couldn't live at all. That's why I'd say that it's better to leave the tree alone, and when it's already there in the middle of a green part of the city, then the tree also has a reason to be there. (girl, 10 years old)

Trees provide oxygen. And so many trees are being chopped down and poisoned by air pollution. And when a few trees happen to grow somewhere where nobody has paid any attention to them, and on an abandoned lot as well, not too close to the street, and have managed to survive, then I'd leave them alone. Enough trees are being destroyed already in tropical rain forests. (girl, 10 years old)

In these comments humans seem to be conceived as a possible threat to nature; thoughts of this kind may even culminate in the extreme view that nature would be better off without humans: "If there weren't any people on earth, there'd be a lot less damage" (girl, 10 years old).

In other cases, however, children seem to favor a naturalistic explanation and simply regard environmental destruction as inevitable. On the one hand, it may be something over which we as humans have no control: "Deer strip the barks off trees, and that isn't very nice, but you have to let nature take its course" (girl, 10 years old).

On the other hand, destruction by humans is perceived as necessary, because we have to use nature in order to survive. Thus some children may argue that there is plenty of nature around, and that humans are nature too: "We humans are nature too . . . rocks, water, sand, that's all nature" (boy, 10 years old).

An argument of this kind is highly ambivalent. Referring to humans as part of nature can also serve as a source of naturalistic justification for their destructive behavior, as the following remarks of 10-year-olds indicate:

"That used to be a meadow, and we've destroyed everything" (girl, 10 years old).

"So what? We did it because we need it. Where else are we supposed to live? In trees?" (first boy, 10 years old).

"Animals kill each other too. We do it in order to survive" (second boy, 10 years old).

"We have to live somehow or other too" (first boy).

Reference to inevitable laws of nature and humans' undeniable need to use nature as a resource is an argumentative construction that apparently can counteract the moral pressure engendered by anthropomorphic interpretations.

Anthropomorphism and Ecosystems

Ecosystems as "wholes" were not spontaneously anthropomorphized by children and adolescents in the discussions we conducted. We found no evidence of personification of an ecosystem as it is commonly found in the stories of North American Indians, the myths of eastern cultures, or the journals of Henry David Thoreau. This may have to do with the ecosystems chosen for debate, which were an anonymous apple orchard, pond, or woods. Perhaps the Elbe River might have been more conducive to anthropomorphic interpretation among children in Hamburg since it is familiar and has distinct boundaries. However, children readily focus on the animals in the ecosystem and lament their pending peril in anthropomorphic terms. This in turn leads them to advocate protecting the surroundings in which the animals live.

Among adolescents, the more prevalent concept of animals and their role in ecosystems is one that seems to oscillate between ecologically informed functionalism and metaphysical teleology, as the following quotations from 15–16-year-olds indicate:

Besides, then another part of the cycle will be lost. Every animal fulfills a certain purpose in a smaller or larger cycle. And if we constantly remove some little parts of them, we'll throw the whole system out of order. And sooner or later we'll suffer too. (boy, 16 years old)

I mean, every animal has a certain, ahm, mission, you know. . . . Every animal is needed in that sense. (boy, 16 years old)

I think that every little insect has a mission, it has to work, some kind [sic], I don't know, collect honey or something, I don't know. So I don't think I have the right to kill, ahm, my garden workers or my house workers. (girl, 16 years old)

The last quotation in particular suggests that anthropomorphic interpretation is involved in this kind of thinking as well.

In discussions about ecosystems, moral pressure to preserve the ecosystem is generated by focusing on the animals in these systems and interpreting them in anthropomorphic terms. If the option of not preserving the ecosystem is particularly attractive, however, other cognitive means are found to counteract this pressure. One very effective way to achieve this seems to be to switch to a metaphoric framework in which the ecosystem is conceived as a machine. A machine is not unique and one is just like another; its parts are interchangeable and humans are master mechanics capable of manipulating it. This kind of thinking is in line with that of modern science and technology and obviously is necessary for managing nature. It pervades everyday experience, and it is therefore no surprise that it can be found in all age groups, as the following solutions proposed in response to a dilemma involving an ecosystem indicate: relocate the animals; replant the trees in a different place; move the entire apple orchard to a different place; create a new apple orchard in a different place as compensation for a destroyed one; remove only half the orchard or allow a few trees to remain.

Very different results were obtained with a story involving ecosystem damage that had already occurred instead of potential damage. The situation to be debated has to do with a family that goes on vacation and drives to a beach, only to find it polluted with oil. In this case spending time in the ecosystem is viewed as a desirable end that cannot be attained because of damage to the ecosystem. In contrast, in the other stories we have used, damage to the ecosystem was presented as a means for obtaining a different desirable end (youth center, disco, playground, etc.). And since adults are responsible for the damage in the story about the oil spill, there is no appeal to personal responsibility on the part of the listeners. When this story is discussed by 10-year-olds, forms of argumentation more reminiscent of a transpersonal sphere of judgment (Eckensberger, 1993) appear. Moral indignation is directed against those responsible for the oil spill, and reference is made to societal organizations such as Greenpeace and government, as well as to protest activities such as demonstrations. This situation may be more similar to those that Kahn (1999) used as a basis for his interviews with children.

Anthropomorphism, Physiomorphism, and Identity

In addition to the widespread use of anthropomorphic interpretations of nonhuman nature, our material also indicates that children refer to certain aspects of nature and natural objects in order to interpret their own experiences or those of other humans, as illustrated by the following quotation: "Weeds are the same as trees. You can also compare weeds with people. Like you might say to a black, 'You're a weed. Piss off!' The way the Nazis might talk to foreigners" (boy, 10 years old).

Since interpreting human experience in terms of nonhuman nature or natural objects is formally the opposite of anthropomorphism, we refer to this process as physiomorphism. It can also be observed in discussions about environmental destruction, in which the topic of death is often brought up (see also Gebhard, 1998). For example, by means of physiomorphic interpretation the death of a tree may remind children of their own death, as the following discussion among 10-year-olds demonstrates:

What happens when something [part of the treehouse] [sic] falls apart? The nails will be left stuck in the tree and a few boards, and then the whole healthy tree will be a mess. (first boy)
   Yeah, then it'll get sick and die. (girl)
   So what? People die too. (second boy)
   You want to live as long as you can too, say for a hundred years. And a tree can become even older. And a tree wants to live as long as it can too and not die ahead of time. (girl)

The concepts of anthropomorphism and physiomorphism are also reflected in two basic metaphorical schemes identified by the cognitive linguists Lakoff and Johnson (1980). In the first case human experience is described in terms of a nonhuman object, as, for example, when human development is described as a pathway through life. This is comparable to a physiomorphic interpretation. The second strategy is comparable to anthropomorphism and involves what the authors call "embodiment," that is, using the body as a source for generating metaphors. Lakoff and Johnson assume that one's own body and concrete sensual experience mediated by it is the starting point for all understanding of the external world.

If anthropomorphic interpretation of nature and physiomorphic interpretation of personal experience are reciprocal facets of one and the same process, then this process must correspond to something like a never-ending series of mirrors reflecting one another (metaphor borrowed from S. Opotow). Thus we may draw upon experiences with natural objects for understanding ourselves (physiomorphism), but in turn our representations of these natural objects will have arisen by interpreting them in terms of ourselves and our personal experience (anthropomorphism).

We know very little about this interpretive circle, but we assume that both processes, anthropomorphism and physiomorphism, are examples of an epistemological pathway that Boesch (1978) refers to as "subjectification." According to Boesch, objects are represented in the mind through an active, constructive process of symbolization in the course of subjectification. However, these symbolic representations are not exact reflections of the external world but are loaded with subjective meaning as well. By attaching subjective meaning to an object, the object and the self become mentally intertwined and a unique relationship between the two is established. It is perhaps in this manner that external objects contribute to the formation of personal identity. The reciprocity between anthropomorphic interpretations of nature and physiomorphic interpretations of self may provide a key to understanding how elaborated concepts of both self and nature can be constructed.

Anthropomorphism and Scientific Views

As discussed earlier, anthropomorphism is an example of an epistemological pathway by which emotional relationships with objects are established that allow them to be invested with individual meaning. This process, however, is contradictory to the program of modern science that predominates in our society. The main goal of science is to generate objective knowledge and produce facts by means that are less contingent upon personal idiosyncrasies. Therefore advocates of modern science have traditionally regarded anthropomorphism with skepticism or even hostility and would probably object to anthropomorphism as an acceptable basis for environmental education.

One argument skeptics might advance is that anthropomorphic interpretation is by no means a foolproof guaranty for the "interests" of nature or natural objects, even if it is emotionally charged. If strong personal interests are at stake, these may override prosocial inclinations induced by anthropomorphic interpretation. Indeed, our investigation has shown, for example, that if the issue at stake is having a real Christmas tree rather than an artificial one, younger children may favor having the tree chopped down, even though they also believe that trees feel pain. There is ample evidence that in cultures in which prescientific thinking colored by anthropomorphism is the primary way of thinking, nature is not always treated respectfully (see, for example, the case of New Guineans discussed by Diamond, 1993). A second source of skepticism is the observation that there are many natural objects that do not lend themselves well to anthropomorphic interpretation—grass, fish, sand dunes, and complex and more abstract objects such as ecosystems and species—an observation supported by our study. These objects cannot be readily grasped by simple analogies to human experience. Furthermore, as scientists may rightfully maintain, to deal with a nonhuman object adequately, we have to take its "otherness" into account, its specific characteristics and "needs." Exploring otherness requires a "form of thought . . . that relies less on the specifics of an individual's make up and position in the world, or on the character or type of creature he is. Reality is progressively revealed . . . [by] gradually detaching from the contingencies of the self" (Nagel, 1986, cited in Hailwood, 2000, p. 5). In other words, objective knowledge based on information from other sources and intersubjectivity are required for a more thorough understanding of otherness.

The quotations in which slightly older children (11–12 years) speculate about a tree's ability to think or speak indicate that they are at least aware of its otherness. And in discussions about organisms other than trees, particularly animals, children occasionally also talk about the specific needs of these organisms, thus exhibiting a kind of thinking that Kahn (1999) refers to as "transmorphic reasoning." Interesting in this connection are also our own observations and those of other studies (see, for example, Mähler, 1995) which indicate that children are capable of switching back and forth between anthropomorphic interpretations of nature and more scientific ones. The two epistemological pathways do not seem to be mutually exclusive in their minds. Thus 10-year-olds converse knowledgeably about photosynthesis and oxygen production, but are also convinced that trees feel pain.

Sometimes anthropomorphic interpretations of the needs of an organism or the functional relationships in an ecosystem may be at odds with scientifically informed ones, and it becomes necessary to choose between them. In other cases, however, it is possible that concern generated by anthropomorphic thought can be complemented by objective knowledge. The examples presented here indicate that an anthropomorphic interpretation of nature is certainly not sufficient for dealing with problems in a society imbued with modern science and technology. However, trying to eliminate it altogether is not only imprudent, as we discuss later it may be virtually impossible, as the work by Lakoff and Johnson (1980) suggests.

Anthropomorphism as a Psychological Foundation for Developing Respect for Nature

The results presented in this chapter suggest that the bond between the self and the nonhuman world expressed in anthropomorphic interpretations is an affective one, and that drawing analogies between humans and nonhuman objects of nature may be a fundamental operation in regarding them as moral objects. If this is true, and some philosophers believe it is (for example, Naess, 1989 and Spaemann, 2000), then it might be worthwhile to more closely consider the psychological state in which anthropomorphism occurs.

When children anthropomorphize a tree, it appears that they classify it as a member of a category that includes both humans and trees. The distinguishing qualities of this category, as our results suggest, are the ability to feel pain and the quality of being alive, qualities to which children readily refer in justifying their defense of the tree. Perhaps this is an example of what Nunner-Winkler (2000) calls categorical identity as opposed to personal identity, although she refers only to humans as sources of categorical identity. One could speculate that categorical identity corresponds to a mental state in which property attributions and projections flow freely between representations of objects of the same category. Indeed, as discussed earlier, anthropomorphic interpretations of nature and physiomorphic interpretations of humans seem to occur coincidentally.

However, at some point in development a child's concept of categorical identity apparently may become altered or others may become superimposed upon it. Maybe this occurs as the differences between trees and humans become more evident in the course of science education, as already discussed. Perhaps a strengthening of personal identity leads to the weakening of a categorical identity that includes trees. Or maybe humans and relationships to humans assume a more and more important role in children's lives so that social identity becomes the prominent form of categorical identity.

In view of these observations, we feel that it would be wise to reconsider the status of anthropomorphism in our society and in education. It is often regarded as a primitive or infantile perspective that had best be overcome. Three things that might contribute to this view can be inferred from our earlier discussion: namely, science education, individualization, and an enhanced orientation toward human needs and interests. However, if anthropomorphism is indeed indicative of a kind of categorical identity that permits nature to be moralized, then it might be something we should nurture rather than eliminate.

Nevertheless, science, individualization, and respect for human needs and interests are obviously important aspects of modern culture. Thus although visions of nature or parts of it as humanlike are very powerful and advantageous for protecting nature, others certainly must be cultivated as well, in particular those required for macromorality or moral judgment in a transpersonal sphere. It follows that we must find ways to deal with the tensions that arise with conflicting world views generated by varying processes of identity formation. Ideally, people should be mentally free and flexible enough to be able to move back and forth between various different views (including anthropomorphism) in reaching a decision on how to deal with a problem involving nature (for a discussion of such "visionary flexibility," see Nevers, Billmann-Mahecha and Gebhard, in press). However, this would require that anthropomorphism be permitted to exist as one of a number of different interpretations of nature.

We believe that group discussions in the style of Philosophy for Children constitute a valuable pedagogical tool for achieving these ends. They can provide a unique and protective setting in which conflicting views, including intuitive anthropomorphic ones, can be expressed and debated without danger of repression or ridicule. Another approach already extensively used in environmental education involves providing experiences in and with nature that encourage interpretations of nature other than mechanistic ones; for example, organizing situations in which aesthetic pleasure or awe are experienced or others in which care for nature or natural objects is required. Finally, it might be possible to transform intuitive anthropomorphism into a more enlightened form through objective knowledge that emphasizes the commonalities that undoubtedly exist between humans and other parts of nature, but at the same time offer a view of the relatively modest status of humans in a greater context of time and space. Planetary and evolutionary history, for example, reveals both the common ancestry of humans and other parts of nature as well as the relatively short period of time and space in which humans have existed compared with the rest of nature. Structural and functional continuities can be demonstrated by other scientific disciplines such as anatomy, physiology, and molecular biology. A combination of these approaches might permit anthropomorphism to be sustained and further cultivated.

Acknowledgment

This work was supported by a grant from the German Research Council (Deutsche Forschungsgemeinschaft, DFG).

References

Boesch, E. E. (1978). Kultur und biotop [Culture and biotope]. In C. F. Graumann (Ed.), *Ökologische perspektiven in der psychologie* [Ecological perspectives in psychology] (pp. 11–32). Bern: Huber.

Callicott, J. B. (1980). Animal liberation: A triangular affair. *Environmental Ethics, 2,* 311–338.

Diamond, J. (1993). New Guineans and their natural world. In S. R. Kellert & E. O. Wilson (Eds.). *The biophilia hypothesis* (pp. 251–271). Washington, D.C.: Island Press.

Eckensberger, L. H. (1993). Normative und deskriptive, strukturelle und empirische anteile in moralischen urteilen: Ein Ökonomie-Ökologie-Konflikt aus psychologischer sicht [Normative and descriptive, structural and empirical aspects of moral judgments. A conflict between economy and ecology from the perspective of psychology]. In L. H. Eckensberger & U. Gähde (Eds.). *Ethische Norm und empirische Hypothese* [Ethical norms and empirical hypotheses] (pp. 328–379). Frankfurt, Germany: Suhrkamp.

Eisenberg, N., Shell, R., Pasternack, J., Lennon R., Beller, R., & Mathy, R. M. (1987). Prosocial development in middle childhood: A longitudinal study. *Developmental Psychology, 23,* 712–718.

Erikson, E. H. (1950). *Childhood and society.* New York: Norton.

Gebhard, U. (1998). Todesverdrängung und umweltzerstörung [Repression of death and destruction of the environment]. In U. Becker, K. Feldmann, & F. Johannsen (Eds.). *Sterben und tod in Europa* [Dying and death in Europe] (pp. 145–158). Neukirchen, Germany: Neukirchener Verlag.

Gebhard, U. (2001). *Kind und natur. Die bedeutung der natur für die psychische entwicklung* [The child and nature. The significance of nature for psychological development], 2nd ed. Opladen, Germany: Westdeutscher Verlag.

Giddens, A. (1991). *Modernity and self-identity.* Cambridge: Polity.

de Haan, G., & Kuckartz, U. (1998). Umweltbewusstseinsforschung und umweltbildungsforschung: Stand, trends, ideen [Research on environmental awareness and environmental education. State of the art, trends, ideas]. In G. de Haan & U. Kuckartz (Eds.), *Umweltbildung und umweltbewusstsein. Forschungsperspektiven im kontext nachhaltiger entwicklung* [Environmental education and environmental awareness. Research perspectives in the context of sustainable development] (pp. 13–38). Opladen, Germany: Leske and Budrich.

Hailwood, S. A. (2000). The value of nature's otherness. *Environmental Values,* 9, 353–372.

Hargrove, E. C. (Ed.) (1992). *The animal rights/environmental ethics debate. The environmental perspective.* Albany: State University of New York Press.

Kahn, P. (1999). *The human relationship with nature.* Cambridge, Mass.: MIT Press.

Lakoff, G., & Johnson, M. (1980). *Metaphors we live by.* Chicago: University of Chicago Press.

Lehmann, J. (1999). *Befunde empirischer forschung zu umweltbildung und umweltbewußtsein* [Results of empirical research on environmental education and environmental awareness]. Opladen, Germany: Leske and Budrich.

Lipman, M., & Sharp, A. M. (Eds.) (1978). *Growing up with philosophy.* Philadelphia: Temple University Press.

Mähler, C. (1995). *Weiß die Sonne, daß sie scheint? Eine experimentelle studie zur deutung des animistischen denkens bei kindern* [Does the sun know that it's shining? An experimental investigation of the significance of animistic thinking in children]. Münster, Germany: Waxmann.

Matthews, G. B. (1984). *Dialogues with children.* Cambridge, Mass.: Harvard University Press.

Mead, G. H. (1934). *Mind, self and society.* Chicago: University of Chicago Press.

Naess, A. (1989). *Ecology, community and lifestyle.* Cambridge: Cambridge University Press.

Nagel, T. (1986). *The view from nowhere.* Oxford: Oxford University Press.

Nevers, P., Billmann-Mahecha, E., & Gebhard, U. (in press). Visions of nature and value orientations among German children and adolescents. In R. van den Born, W. de Groot, & R. Lenders (Eds.), *Visions of nature. A research-based exploration of people's implicit philosophies.* Dordrecht: Kluwer.

Nevers, P., Gebhard, U., & Billmann, E. (1997). Patterns of reasoning exhibited by children and adolescents in response to moral dilemmas involving plants, animals and ecosystems. *Journal of Moral Education,* 26, 169–186.

Nunner-Winkler, G. (2000). Identität aus soziologischer sicht [Identity from the perspective of sociology]. In W. Greve (Ed.). *Psychologie des selbst* [The psychology of the self] (pp. 302–316). Weinheim, Germany: Psychologie Verlags Union.

Opotow, S. (1987). Limits of fairnesss: An experimental examination of antecedents of the scope of justice. Doctoral dissertation, Columbia University.

Opotow, S. (1994). Predicting protection: Scope of justice and the natural world. *Journal of Social Issues, 50*, 49–63.

Rest, J., Narvaez, N., Bebeau, M. J., & Thoma, S. J. (1999). *Postconventional moral thinking. A neo-Kohlbergian approach.* Mahwah, N.J.: Lawrence Erlbaum.

Schweitzer, A.(1995). [Original: 1980, Felix Meiner Verlag, Hamburg, Germany] *Aus meinem Leben und Denken* [About my life and my thoughts]. Frankfurt, Germany: Fischer Verlag.

Searles, H. F. (1960). *The nonhuman environment in normal development and schizophrenia.* New York: International Universities Press.

Spaemann, R. (2000). Wirklichkeit als anthropomorphismus [Reality as anthropomorphism]. *Information Philosophie, 4*, 7–18.

Strauss, A., & Corbin, J. (1996). *Grounded theory. Grundlagen qualitativer sozialforschung* [Grounded theory. Basics of qualitative research in the social sciences]. Weinheim, Germany: Psychologie Verlags Union.

Taylor, P. (1986). *Respect for nature.* Princeton, N.J.: Princeton University Press.

6

The Development of Environmental Moral Identity

Peter H. Kahn, Jr.

One day I met my neighbor, a logger, on a dusty dirt road near our respective lands, an hour's drive from the nearest small town. His name is Horse. He is a big fellow, part White and part Indian. On that day he tells me that he is heading 5 miles north to where he and his crew are logging on a 30,000-acre cattle ranch. He adds loudly: "Now I ain't hurtin' the environment any. You know, I love this land." I do know. And I knew his dad, too, who had been foreman of another large cattle ranch 10 miles east. Decades earlier, as an adolescent, I had on more than one occasion ridden horseback through that land—trespassed if truth be told. Horse and I came of age in these mountains. I answered Horse back, "Heck, the trees you're cutting, they're mostly overly mature trees, don't you think?" And he looks pleased that I had remembered the point he made during a conversation last year: that he's harvesting trees that are soon to die anyway and so he's doing no harm. "Horse, you know, I'm starting to feel a bit overly mature myself. I hope no one starts a comin' after me." "Oh hooo," Horse bellows, and he drives north. We were glad to see each other.

But I remain puzzled. How do people whose identities appear so deeply connected to the land they love engage in environmentally harmful activities? Do they really believe that the activity (e.g., logging mature trees) causes no environmental harm? Do economic demands simply override environmental moral judgments? Or do both the demands and judgments coexist in an uneasy if unequal alliance? What does it mean when such people say that they love the land? Do they love the land only for what it can give to them, or in some way that extends beyond their own immediate self-interest?

Such questions have formed part of my research, which aims toward an account of the human relationship with nature (Kahn, 1994, 1997a,b; 1999; Kahn & Friedman, 1995, 1998; Kahn & Kellert, 2002). To succeed, or course, any such account must be large in scope and inter-disciplinary. It would include, for example, investigations into our evolutionary history within a Darwinian framework: that certain responses to nature have been more adaptive than others (e.g., fear of snakes or an attraction to bodies of water) and thus persist in who we are today (Kellert, 1997; Kellert & Wilson, 1993; Wilson, 1975, 1984). It would also include social, political, and historical investigations (e.g., Berry, 1997; Nash, 1973; Orr, 1994). My own focus over the years—while attentive to such investigations—has been more on understanding the development of children's environmental moral reasoning and values. Such understandings, when attained, capture that which is at once deeply fundamental to our being and very practical. In his classic essay on the conservation ethic, Aldo Leopold (1970) argues that environmental education will continue to fail until we help people develop a "love, respect, and admiration for land, and a high regard for its value" (p. 261). "No important change in ethics," Leopold writes, "was ever accomplished without an internal change in our intellectual emphasis, loyalties, affections, and convictions" (p. 246). That is much of what I have been seeking: understanding, with respect to nature, children's intellectual emphasis, loyalties, affections, and convictions.

Thus I have two goals in this chapter. First, I present some of my central findings on children's environmental moral reasoning. I have presented some of this material elsewhere. I also take this occasion to raise a particularly puzzling aspect of the cross-cultural data on the relative effects of development and culture in forming an individual's environmental conceptions. Second, I show how this research and theorizing bear on understanding the child's construction of environmental moral identity. I suggest that through structural development the multiplicities of environmental identity are integrated and transformed into a largely coherent and unified sense of self.

Environmental Moral Reasoning

To convey a sense of children's environmental moral reasoning, I draw selectively from five collaborative studies. Two of them involved a black population in an economically impoverished community in Houston, Texas. In the first study (which will be referred to as the Houston child study) we interviewed 72 children evenly divided across grades 1, 3, and 5 (Kahn & Friedman, 1995). In the second study (the Houston parent study) we interviewed 24 parents from the school that participated in the Houston child study (Kahn & Friedman, 1998). In the third study (the Prince William Sound study) we interviewed 60 children in Houston across grades 2, 5, and 8, on their moral and ecological reasoning about the 1990 *Exxon-Valdez* oil spill that occurred in Prince William Sound, Alaska (Kahn, 1997b). In the fourth study (the Amazonia study) we modified the methods from the Houston child study and interviewed in Portuguese 44 fifth-grade Brazilian children in urban and rural parts of the Amazon jungle (Howe, Kahn, & Friedman, 1996). In the fifth study (the Lisboa study) we interviewed in Portuguese 120 children and young adults in grades 5, 8, 11, and college in Lisbon, Portugal (Kahn & Lourenço, 2002).

Methodologically, we employed the semistructured interview pioneered by Piaget (e.g., 1960, 1969, 1983) which has been elaborated by a large number of more recent researchers (e.g., Damon, 1977; Ginsburg, 1997; Helwig, 1995; Killen, 1990; Kohlberg, 1984; Lourenço, 1990; Smetana, 1995; Turiel, 1983). Some of our interview questions focused on the pollution of a local waterway: a nearby bayou in Houston; the Rio Negro in Amazonia; and the Rio Tejo in Lisboa. Other questions focused on participants' (1) environmental commitments and practices; (2) moral understandings about human actions that affect such everyday natural phenomena as birds, water, plants, insects, open spaces, and air; (3) potentially contradictory environmental judgments; (4) conceptions of what counts as "natural" activity; and (5) conceptions of what it means to live in harmony with nature.

I start this discussion with some results from the Houston child study. Keep in mind that the children we interviewed came from one of the most economically impoverished communities in Houston (which I will say more about later). Of the children we interviewed, the majority (84 percent) said that animals played an important part in their lives, as did plants (87 percent) and parks or open spaces (70 percent). The majority of the children (72 percent) talked about environmental issues (such as pollution) with their family, and did things to help the environment, such as recycling (74 percent) or picking up garbage (25 percent).

Children judged that polluting a bayou would have harmful effects on birds (94 percent), water (91 percent), insects (77 percent), and the view (93 percent). Moreover, it is one thing to know that harm is occurring to an entity, it is another thing to care that that harm is going on. Our results showed that it would matter to these children if such harm occurred to birds (89 percent), water (91 percent), insects (77 percent), and the view (93 percent).

We also analyzed whether children judged the act of throwing garbage in their local bayou as a violation of a moral obligation. We drew here on the domain literature of Turiel (1983, 1998), Nucci (1981, 1996), Smetana (1983, 1995), and others in which a moral obligation is assessed, in part, based on the criterion judgments of prescriptivity (e.g., it is not all right to throw garbage in a bayou), rule contingency (the act is not all right even if the law says it is), and generalizability (the act is not all right for people in another country, even if those people perform the act). Based on these and three other criterion judgments, and in conjunction with children's moral justifications, our results showed that the majority of the children believed it was morally obligatory not to throw garbage in a bayou. Developmentally, fewer children in grade 1 (68 percent) than in grades 3 (91 percent) and 5 (100 percent) provided such morally obligatory judgments.

In this study we also characterized children's environmental moral reasoning. In the broadest perspective, two main forms of environmental reasoning emerged from the data: anthropocentric and biocentric. Anthropocentric reasoning appeals to how effects on the environment affect human beings. Justification categories included appeals to (1) personal interests (e.g., "animals matter to me a little bit because we need more pets and different animals to play with"); (2) aesthetics (e.g., "because I'd get to see all the colors of the plants and the beauty of the whole—of the whole natural plants"); and (3) the physical, material, and psychological welfare of self and others (e.g., "air pollution goes by and people get sick, it really bothers me because that could be another person's life"). In turn, biocentric reasoning appeals to a larger ecological community of which humans may be a part. The justification categories included appeals to the intrinsic value of nature (e.g., "if nature made birds, nature does not want to see birds die") and to the rights of nature or the concept that nature deserves respect (e.g., "they [animals] need the same respect we need").

Isomorphic and Transmorphic Reasoning

Two ways in particular emerged from the data for how children established reasoning on biocentric rights. One way occurred through establishing isomorphic relationships. Here natural entities (usually animals) were compared directly with humans. For example, one child said: "Fishes, they want to live freely, just like we live freely. . . . They have to live in freedom, because they don't like living in an environment where there is [so] much pollution that they die every day." Thus an animal's desire ("to live freely") is viewed as equivalent to that of a human's desire, and because of this direct equivalence, children reasoned that animals merit the same moral consideration as humans. A second way occurred through establishing transmorphic relationships. For example, a fifth-grade child said:

Fish need the same respect as we need. . . . Fishes don't have the same things we have. But they do the same things. They don't have noses, but they have scales to breathe, and they have mouths like we have mouths. And they have eyes like we have eyes. And they have the same co-ordinates we have. . . . A co-ordinate is something like, if you have something different, then I'm going to have something, but it's going to be the same. Just going to be different.

This child appears to draw on a word he encountered in some other context to help him explain that while fish are in some respects not the same as people (they do not have noses like those of people), in important functions (such as breathing and seeing) they are the same. Thus he moves beyond a reciprocity based on directly perceivable and salient characteristics to establish moral equivalences based on functional properties.

Isomorphic and transmorphic reasoning should not be confused with anthropomorphic reasoning. In the latter case, an aspect of nature is thought to have human properties. Consider, for example, the anthropomorphic reasoning of the following child from the Amazonia study (translated from Portuguese) as she was explaining why the government should stop people from logging the jungle:

It is like me having a leg or an arm cut. . . . Nature is like a person, no, thousands of persons because it isn't just one thing. . . . [A] person is like a tree. If the tree bears fruits, it is the same with people. Taking care of a tree is the same. If you cut a branch off a tree it is like cutting a finger or the foot. To cut a tree down is like doing it to yourself. It is the same to our heart, it is not good. The jungle is like the heart of a person.

Here nature is likened to a human or becomes human in one or more important ways. In contrast, as noted earlier, in isomorphic reasoning a moral feature (such as freedom) is deemed important to both nature and humans, and on that basis a moral principle (such as protecting freedom) is applied equally to both nature and humans ("Fishes, they want to live freely, just like we live freely").

Developmentally, the child's understanding of animals appears to start early in childhood. Myers (1998; Myers & Saunders, 2002), for example, provides evidence that even by 3 months of age children begin to develop understandings that animals display four properties that remain constant across many different interactions: *agency* (a dog decides to eat and acts accordingly), *affectivity* (a dog appears to enjoy playing with the child), *coherence* (a dog is able to coordinate its movements in response to the child's actions), and *continuity* (the dog's repeated interactions become regularized into a relationship with the child). Such understandings, according to Myers, make it possible for children to recognize that animals have their own subjective states and can have correlative interests in interacting with the child ("my dog wants to play with me"). These cognitive underpinnings, in turn, make possible the development of caring for individual animals. Such caring, however, can be limited. After all, what about animals children do not know personally: a dog across town? macaques in Indonesia? Presumably such animals also deserve moral consideration. Thus my account of the development of children's isomorphic and transmorphic biocentric moral reasoning extends Myers's theorizing insofar as it characterizes increasingly complex levels of moral reasoning that allow older children to construct generalized concepts of care—for animals in general and potentially the natural world as a whole.

Isomorphic and transmorphic reasoning may also provide the developmental underpinnings for yet another evolutionarily shaped relationship with nature. What I have in mind is this. In his book *The Others: How Animals Made Us Human*, Shepard (1996) argues that animals were "among the first objects of classificatory thinking" (p. 97) and that "the human species emerged enacting, dreaming, and thinking animals and cannot be fully itself without them" (p. 4). "Of each species," Shepard proposes, "we can say, 'I am not that—and yet, just in this one respect, it is like a part of me,' and so on, as though with every 'I am not that one' we keep some bit of them. We take in the animal, disgorge part of it, discover who we are and are not" (p. 72). Thus, in comparison with anthropomorphic reasoning, isomorphic reasoning—and to a larger extent transmorphic reasoning—grants greater independence to the natural world, embracing what Shepard (1996) refers to as "otherness": the partly unknown and wild aspects of nature that "is essential to the discovery of the true self" (p. 5).

Cross-Cultural Similarities

One of the striking features across our five studies was the degree of cross-cultural similarity. For example, by and large participants across studies said that animals, plants, and open spaces played an important part in their lives, were aware of environmental problems, recognized that pollution harmed various natural entities (e.g., birds, water, insects, and the view), would care if such harm occurred, and brought moral obligatory reasoning (based on the criteria of prescriptivity, noncontingency of conventional practices, and generalizability) to their environmental judgments.

Even more striking, perhaps, was the degree to which the participants' reasoning seemed virtually identical across locations. To provide a sense of the substance here—that the similarities do not merely reflect superficial resemblances—I provide some matched examples of reasoning across locations. I start with an example that speaks to a common if not visceral response people have to a polluted environment:

1A. [The people by the river would be affected because] the smell of the water, it should bother people to open their windows and feel that foul smell. . . . [It would matter to me] because a person shouldn't have to smell dead fish or trash bags full of rotten stuff when she opens the window in the morning. (Lisboa study)

1B. [The air] stinks, 'cause I laid up in the bed the other night. Kept smelling something, knew it wasn't in my house, 'cause I try to keep everything clean. Went to the window and it almost knocked me out. The scent was coming from outdoors into the inside and I didn't know where it was coming from. . . . Now, who'd want to walk around smelling that all the time? (Houston parent study)

Such reasoning grounded the participants' judgments that it is wrong to throw garbage in a local waterway or to pollute the air.

Other examples speak to the importance of trees in the healthy psychological functioning of human lives:

2A. I live in the country and I find that living in the city is very difficult, it causes stress. For instance, we live on this street full of trees. Anytime that I leave home in the morning, I feel invigorated seeing the trees and their shade, I can breathe, I can hear the birds. Now, if I lived on a street close to Avenida da Republica, I would feel stressed seeing that amount of cars, very few trees. (Lisboa study)

2B. Yesterday, as my son and I were walking to the store and we were walking down Alabama [street] and for some reason, I think they're getting ready to widen the street. And it's a section of Alabama that I thought was so beautiful because of the trees and they've cut down all the trees. And you know it hurts me every time I walk that way and I hadn't realized that my son had paid attention to it, too. So, he asked me, he said, "Mama, why are these, why have they cut down all the trees?" And then he asked me, "Well, if they cut down all the trees everywhere, would that have an effect on how we breathe?" (Houston parent study)

Thus I do not think it is the case that aspects of everyday nature—trees, plants, open spaces, sunshine, fresh air—are luxuries of the well-to-do; rather, they are psychological necessities that people often recognize. I will expand on this point later.

I provided earlier an example, from the Houston child study. Consider now two virtually identical examples from children in Amazonia and Lisboa:

3A. Even if the animals are not human beings, for them they are the same as we are, they think like we do. (Amazonia study)

3B. [Wild animals are important] because they breathe like we do, and sometimes we think that because they are animals they are not like us, that they don't do certain things. Then we end up seeing that they do. (Lisboa study)

Aristotle (1962) begins his *Nicomachean Ethics* by saying that "the good, therefore, has been well defined as that at which all things aim." He then develops a teleological account of the good in which each kind of inanimate object (e.g., a clock) and animate being (e.g., a human) has an ideal way of functioning. Something of this Aristotelian orientation emerged from the study data. Consider examples across four studies:

4A. Yea, because it looks better. . . . Well, I mean without any animals the world is like incomplete, it's like a paper that's not finished. (Prince William Sound study)

4B. Because water is what nature made; nature didn't make water to be purple and stuff like that, just one color. When you're dealing with what nature made, you need not destroy it. (Houston child study)

4C. Because the river was not made to have trash thrown in it, because the river belongs to nature. (Amazonia study)

4D. [Wild animals] are important because if someone created them it is because they have some kind of role. (Lisboa study)

All these participants offer a moral conception of the proper endpoint of nature, and that good arises from nature reaching that end and being complete.

By emphasizing cross-cultural similarities, I do not want to run roughshod over the unique features of each cultural context and how these features, too, can shape children's environmental commitments and sensibilities. When we asked participants in the Houston child study what they thought about in terms of nature, for example, 7 percent of the children responded with issues pertaining to drugs and human violence; and when asked about what environmental issues they talk about with their families, 17 percent of the children responded with similar issues. As one fifth-grade girl said when she was merely describing a bayou (a preliminary question in the interview):

It's where turtles live and the water is green because it is polluted. People—some people need to um, some people are nasty. Some people, you know, like some people go down there and pee in the water. MM HMM. Like boys, they don't have no where to pee, and drunkers, they'll go do that, too. OKAY. And sometimes they'll take people down and rape them, and when they finished, they might throw 'em in the water or something. SO, WHAT DOES IT LOOK LIKE? HOW WOULD YOU DESCRIBE IT? A BAYOU? It's big and long and green and it stinks . . . And turtles live in it.

Indeed, it was this element of human violence and danger that often prevented children in Houston from experiencing nature. For example, consider one of the first-grade children in this study, Eboni, who seemed to us the least connected to or interested in nature. We asked her, "Do you ever climb trees?" She said no. We asked why. She responded: "Cause it's dangerous. Cause if they fall the grass might have glass and then they fall on they face in the glass and then they'll cut their nose or eyes and they they'll be blind." Eboni told us that she never goes to the local park. We asked why. "Because I used to go, now the people go in there and they be throwing glass and they have guns and stuff and they might shoot me." Indeed, Eboni does not even like to play in her back yard. Why? "Nothin' can get me [indoors]. Like a stranger or something." Thus it is less the case that Eboni has no affiliation with animals, plants, and parks or open spaces, and more that her economically impoverished and violent urban surroundings have made nature largely inaccessible.

The Development of Biocentric Reasoning

In the account of environmental moral reasoning given here, I pay heed to culture. Yet some readers may be thinking: "Well, in all that has been said so far, culture really plays a secondary role, attenuating or modifying what are proposed as more fundamental psychological processes and constructions. Are there not ways that culture plays a primary role?" Perhaps. What I would like to explore now is a puzzling aspect of my data, and see whether a cultural explanation can be invoked to solve it.

The puzzle emerged in the following way: Earlier I showed that both anthropocentric and biocentric reasoning emerged in the Houston child study. However, taken across nine questions that we analyzed systematically, only about 4 percent of the children's reasoning was biocentric. Based on this result, I wondered what we would find if we interviewed children who grew up, not in an inner city, but in a rural village and who lived in daily, intimate connection with the land. Would we, for example, find a greater proportion of biocentric reasoning? This question was one of several that motivated the Amazonia study. Our results showed, however, that even in a small rural village along the Rio Negro (accessible only by boat), there was no statistical difference in the percentage of biocentric reasoning (about 4 percent in Houston, 4 percent in Manaus, and 8 percent in the remote village).

In interpreting these results, I was inclined to believe that across cultures biocentric reasoning emerges more fully in older adolescents and adults. Thus, conducting our next study in Lisboa seemed ideal because we were able to control not only for language (interviewing in Portuguese), but within the very country that had colonized much of Brazil. Thus, if biocentric forms of reasoning were found to increase with age across our Portuguese population, it would provide evidence to support the developmental hypothesis.

Our results showed that on a few questions (which we had not asked in the previous studies) biocentric reasoning increased somewhat with age. For example, we asked "Are wild animals important to you, and why or why not?" On this question, the results showed the following use of biocentric reasoning: fifth grade, 60 percent of justifications were biocentric; eighth grade, 70 percent; eleventh grade, 83 percent; and college, 82 percent). Still, such biocentric orientations were not pervasive. Across most of the eleven questions from which we systematically collected justification data, the proportion of biocentric reasoning in the older population in Portugal roughly matched the small proportion of biocentric reasoning in the Houston child study and the Amazonia study.

In interpreting these results, I had suggested elsewhere (Kahn, 1999) that there was some qualified support for the developmental hypothesis. I said, for example, that perhaps biocentric reasoning has taken shape structurally by adolescence, but then gets employed only occasionally, depending on the context. As an analogy, imagine if you had a sports car that had the capability of going 110 mph. Occasionally you might exercise this capability, but usually driving in the city prevents such activity. Similarly, adolescents and young adults may have developed the capability to engage in biocentric reasoning, but rarely do so.

Such an explanation, however, may misconstrue key ideas that lie at the intersection of biocentric reasoning, development, and culture. Throughout my investigations, I have assumed that there is a single pathway by which biocentric reasoning emerges. But perhaps two pathways actually exist. One pathway may emerge (but for some reason—as shown by the Amazonia study—not in all cases) in cultures that live in daily, intimate contact with the land. Thus, for example, Nelson (1989) reports on the biocentric relationship that the Koyukon of northern Alaska have with their community of nature; a community that includes not only humans, animals, and plants, but also mountains, rivers, lakes, and storms—the Earth itself. As Nelson (1989) writes: "According to Koyukon teachers, the tree I lean against feels me, hears what I say about it, and engages me in a moral reciprocity based on responsible use. In their tradition, the forest is both a provider and a community of spiritually empowered beings. There is no emptiness in the forest, no unwatched solitude, no wilderness where a person moves outside moral judgment and law" (p. 13).

In turn, a second pathway by which a culture can develop a biocentric orientation may depend less on daily, intimate contact with the land, and more on modern philosophical moral discourse. Here there is some historical indication that such moral discourse leads to extending moral standing to an ever-widening range of entities. For example, over the past 150 years in the United States, moral rights have accrued to Blacks, women, and children; and some argue it is just a matter of time before they accrue to animals and nature in general. As Stone (1986) writes: "Each time there is a movement to confer rights onto some new 'entity,' the proposal is bound to sound odd or frightening or laughable. This is partly because until the rightless thing receives its rights, we cannot see it as anything but a thing for the use of 'us'—those who are holding rights at the time. . . . I am quite seriously proposing that we give legal rights to forests, oceans, rivers and other so-called 'natural objects' in the environment—indeed, to the natural environment as a whole" (pp. 84–85).

If this "dual-pathway" account has merit, then are the biocentric conceptions that emerge by means of these two pathways the same? I do not know. To answer this question, one would need psychological research with children, adolescents, and adults in native cultures, such as perhaps the Koyukon, that have a clear biocentric ethos. A major difficulty of conducting such research, of course, is that such cultures are disappearing quickly, absorbed by increased globalization, and are being changed—apparently irreversibly—by the introduction of advanced technologies and western consumer desires.

Environmental Moral Identity

The term *identity* seems fraught with difficulties (Holland, 1997; Shweder & Bourne, 1984; Thoits & Virshup, 1997; McAdams, 1997). As Ashmore and Jussim (1997) write: "Self and identity are not simple concepts. . . . These words point to large, amorphous, and changing phenomena that defy hard and fast definitions" (p. 5). Indeed, one of the most difficult problems for scholars of identity is reconciling two opposing ideas—whether identity as a construct is multiple or unified. On the side of multiplicity is the obvious fact that each one of us has many roles. A single person can be, for example, a father, lover, poker player, gourmet cook, and accountant. Indeed, William James says that a person has as "many social selves as there are individuals who recognize him" (quoted in Rosenberg, 1997, p. 23). On the side of unity, virtually all of us feel as if our sense of identity remains reasonably stable over time. The studies on children's environmental moral reasoning discussed here help explicate these two seemingly contradictory conceptions of identity.

Identity as Multiplicity

The children we studied provided a wide range of environmental justifications, including those based on human welfare, personal interests, landscape aesthetics, the intrinsic value of nature, and the rights of animals. Each of these justifications themselves took various forms which could be coded, and these forms in turn had nuances that could be captured qualitatively. For example, considerations for human welfare can involve three forms. One form is *physical welfare*: "[It's wrong to throw garbage in the bayou] because if the water is dirty, I might get sick"— the Houston child study. A second is *material welfare*: "[It's wrong to throw garbage in the water because] from an economic point of view the water would be captured and sent to a central plant where it would be treated. Who is paying for the process to clean the water? Isn't it us? So, we are causing harm to ourselves"—the Lisboa study. A third form is *psychological welfare*: "[Gardens are important] because the city is a place that causes great stress and it gives a chance to someone to go to a place that is near, and to be in contact with nature, to stay calm"— the Lisboa study.

As shown earlier, sometimes in response to a single question children drew on multiple justifications. Other times different questions elicited one form of reasoning rather than another. The upshot is that even during the early years children do not have a monolithic orientation in reasoning about the natural world.

Different reasons, of course, can justify different courses of action. In such cases, how do children choose among seemingly contradictory considerations? Our Lisboa data bear directly on this question. In that study, we asked the participants whether air pollution was a problem in Lisboa. Ninety-eight percent of them said yes. Then we asked whether driving a car increases the air pollution. One hundred percent of the participants said yes. In this context, we then asked a pivotal question: "Do you think it is all right or not all right that a person drives his or her own car to work every day?" For those participants who said that it was all right, we then countered with the probe: "But how is it all right to drive the car if, as you said before, that increases the air pollution?" For those participants who said it was not all right, we then countered with the probe: "But how could this person arrive at his or her place of work? Would that be practical?" The results showed that in some form or another, 81 percent of the participants believed it was all right for a person to drive his or her car to work. But the reason I say "in some form or another" is that in various ways the participants often qualified their evaluations and sought to coordinate their judgments about pollution with other personal and moral considerations of import. Specifically, we were able to ascertain three overarching forms by which the participants coordinated their judgments concerning the air pollution caused by driving a car with the permissibility of driving: overriding, contextual, and contradictory.

In an *overriding coordination*, one consideration simply overrides other considerations. For example, in supporting a judgment that driving a car is fundamentally not justified, one participant said: "I think that is totally not all right. Because I think that in Lisboa there is good public transportation . . . that comes at reasonable frequency and that is not expensive." However, it is equally possible to use an overriding coordination to argue that driving a car is fundamentally permissible. For example, another participant said: "I think that it is right. Because one needs this asset to go to work, so he won't have to face long lines, like the ones for the buses, so he won't waste so much time." Although these two evaluations differ, the nature of the coordination is the same insofar as each participant upholds a single generalized position. In a *contradictory coordination*, contradictory positions are upheld. For example, one participant said: "It's right because there are a lot of people who don't have public transportation to go to their jobs. . . . Well, it's a contradiction, but it is that way." This participant says "it's a contradiction" because she had just established that the action in question was not all right. In a *contextual coordination*, the judgment depends on the specific context. For example, one participant said: "It depends. If the place of work is very far away and there is no other way of transportation, then one has to take [one's car]. But if there are other ways of transportation that cause less pollution, I think that people should go [that way]. . . . One could also go by bicycle, that helps exercise and doesn't cause pollution."

I have been suggesting that environmental identity embraces aspects of multiplicity by the very diversity of reasoning that children and young adults across different cultures bring to environmental issues. By implication, my perspective (and supporting data) speaks against stereotyping a person or a people. I do not think it can be said, for example, that "loggers are all this way" or that "Greenpeace activists are all another way" (see Kempton, Boster, & Hartley, 1995). Of course, this point may seem rather obvious to the reader. However, a more subtle form of stereotyping occurs in both popular and academic discourse.

As a case in point, consider Inglehart's (1995) study of data from the 1990–93 World Values survey carried out in forty-three countries. From his analysis, Inglehart argued that public concern for environmental quality is stronger in wealthy nations than in poor nations, and that environmentalism is a product of a "postmaterialist culture shift." Inglehart's argument in effect follows Maslow's (1975) theory of a hierarchy of needs: that "someone whose needs for food, shelter and physical security are barely met is not likely to spare the energy—physical or emotional—to maintain concern about [environmental issues]" (Hershey and Hill, 1977–78, quoted in Mohai, 1990, p. 747).

Inglehart's study set off vigorous debate in the political science and sociological literature (e.g., Kidd & Lee, 1997; Martínez-Alier, 1995; Pierce, 1997). I want to highlight one response in particular. Namely, Brechin (1999) sought to refute Inglehart's claim by conducting a study on citizen attitudes toward the environment from twelve relatively wealthy countries and twelve relatively poor ones. Brechin found no statistical differences between the two groups on questions concerning symbolic global environmental problems. For example, he found a "biocentric-postmaterialist concern for other living species" across both groups (p. 807). Most generally, Brechin found no support for the postmaterialist claims, which he views as incoherent when applied cross-culturally on the national level (see Brechin & Kempton, 1994).

Brechin thus brings to bear on the sociological level the very idea I have sought to develop on the psychological level: that even in economically impoverished communities people have a rich and diverse appreciation for nature and a moral responsiveness to its well-being. Granted, such moral regard for nature may not take forms commonly associated with "environmentalism"—such as high regard for wilderness areas—although it might. However, such moral regard for nature extends well beyond a mere interest in combating threats of pollution to one's local community.

Identity as Unity

While children employ rich and diverse forms of reasoning about the natural environment, they also hold to environmental ideas that may change little over time. Drawing on the moral developmental literature (Turiel, 1998), two such ideas can be characterized in terms of assumptions of how the world works: ecological assumptions and metaphysical assumptions.

I sketched an ecological assumption in the opening vignette about my neighbor Horse and his view that logging overly mature trees causes no environmental harm. Similar assumptions emerged from the Lisboa data. For example, consider two responses from college participants:

> To cut [trees] for the sake of cutting, no. To cut with a purpose, and using it in a way that is profitable to people over there . . . is also correct. But . . . instead of cutting down any tree, we cut only the ones that are mature, fully grown. That way it wouldn't be a big problem.

> However, sometimes [cutting trees down] is not deforestation itself. I am talking about cutting trees down, it can be good for the forest, if there are small trees and such, if there is management of that forest, for sure the forest is going to get better. The little ones can't grow because the big ones cast a shadow [over them].

Like Horse, these participants assume that the health and integrity of ecological systems (and specifically forests) do not depend on the existence of mature trees (e.g., for genetic stock) and their eventual death and decay.

Many other such assumptions pervade our societal discourse about environmental issues. Some people assume, for example, that carbon dioxide emissions contribute to global warming; others do not. Some people assume that drilling for oil in the Artic National Wildlife Refuge will cause environmental harm; others do not. Some people assume that agriculture based on genetically altered plants will harm farmlands and ecosystems; others do not. Because ecological assumptions are, at least in principle, open to empirical validation, they provide a means by which environmental science education can change aspects of individuals' environmental identity.

The same cannot be said so readily about metaphysical assumptions. The basic idea here has been worked out well in a continuing debate in the moral development literature. On the one side, Shweder, Mahapatra, and Miller (1987) have argued that there are divergent if not incommensurable moral perspectives in Hindu and Western cultures. To make their case, Shweder et al. have drawn from their extensive interviews with devout Hindus in India. For example, they found that devout Hindus believe it is immoral for a widow to eat fish; and surely (Shweder and his colleagues argue) few people in the United States hold a similar view. On the other side, Turiel, Killen, and Helwig (1987) have argued that what might appear as moral diversity between cultures often involves differences on nonmoral dimensions. For example, in their reinterpretation of Shweder et al.'s example, Turiel et al. found that devout Hindus believed that harmful consequences would follow if a widow ate fish (the act would offend her husband's spirit and cause the widow to suffer greatly). While such metaphysical assumptions—that spirits exist and can be offended by earthly activity—may differ from those in Western culture, the underlying moral concern for the welfare of others appears congruent across cultures. (For a discussion of these issues, see Kahn, 1991; Kahn & Lourenço, 1999; Turiel, 1998, 2002; Wainryb, 1993, 1995).

In my own research I have not systematically sought to elicit individuals' metaphysical assumptions about the environment. Nonetheless, on occasion such assumptions have emerged in the course of my interviews with children. Sometimes these assumptions came almost verbatim from biblical texts. For example, one child from the Houston study offered this response to why he did not eat ham: "[Because] God took the demon out of the man and put the demon in the pig, and that's why since then I never ate ham or anything like that." Other times the metaphysical assumptions were embedded in teleological reasons for why nature has intrinsic value. For example, one child from the Lisboa study argued that wild animals "are important because if someone created them it is because they have some kind of role." Notice that unlike ecological assumptions, metaphysical assumptions are not open to empirical validation, at least in ways understood by western science. It is not easy, after all, to prove or disprove that a creator exists, let alone to divine the Divine's intention.

Conclusion

I started this chapter by asking how people whose identities appear so deeply connected to the land they love can engage in environmentally harmful activities. My answer has been built from a seemingly contradictory conception of identity: that it is at once multiple and unified.

In terms of its multiplicity, I have shown how children bring diverse considerations into their environmental reasoning, including considerations based on personal interests, human welfare, aesthetics, teleology, intrinsic value of nature, and rights and respect for nature. Children also can embrace competing if not contradictory claims (including those that I have characterized as overriding, contradictory, and contextual). In terms of its unity, I have shown how children employ coherent orientations or constellations of orientations across a diverse range of environmental topics. In addition, children across diverse cultures appear to engage in remarkably similar environmental moral reasoning.

From the standpoint of developmental psychological theory, does it make sense to talk about environmental identity in terms of both multiplicity and unity? I think so. When a rock falls on a child's head, the culture does not matter; it physically hurts the child, and from such interactions children construct notions of physical causality. Or if a child from any culture pulls on a dog's tail hard enough, the dog will object (run away, bark, or even bite), and from such interactions children construct notions of animal welfare. These are but simple examples to illustrate a long-recognized structural-developmental mechanism that operates on both the microgenetic and macrogenetic level (Baldwin, 1973). Through interaction with nature, artifacts, and other people, children construct increasingly adequate understandings of the world. Moreover, through the course of child development, earlier forms of knowledge do not so often disappear as they are reworked and transformed into more sophisticated structures. Recall, for example, that transmorphic reasoning subsumes the sophistication of an isomorphic perspective (which finds similarities between humans and animals), but then does something more with it (coordinates similarities with differences). My point is that the underlying cohesion in cognitive structures helps create the sense of unity in identity at any given developmental point in time, while the transformations of the cognitive structures help create our sense of a changing identity over time.

Finally, I would like to return to one of the central findings from my collaborative research: that children across diverse cultures engage in remarkably similar environmental moral reasoning. One explanation for this finding—as I alluded to earlier with the illustration of the falling rock—is that there are universal and invariant aspects of nature itself that give rise to and bound children's environmental constructions. If this explanation is correct, then it follows that while culture plays a fundamental role in the construction of an individual's environmental identity, it is not an unbounded role. Indeed, perhaps it plays less of a role than cultural theorists might suggest.

References

Aristotle (1962). *Nicomachean ethics* (M. Ostwald, Trans.). Indianapolis, Ind.: Bobbs-Merrill.

Ashmore, R. D., & Jussim, L. (1997). Introduction: Toward a second century of the scientific analysis of self and identity. In R. D. Ashmore & L. Jussim (Eds.), *Self and identity* (pp. 3–19). New York: Oxford University Press.

Baldwin, J. M. (1973). *Social and ethical interpretations in mental development.* New York: Arno. (Original work published 1897)

Berry, W. (1977). *The unsettling of America: Culture and agriculture.* New York: Avon.

Brechin, S. R. (1999). Objective problems, subjective values, and global environmentalism: Evaluating the postmaterialist argument and challenging a new explanation. *Social Science Quarterly, 80,* 793–809.

Brechin, S. R., & Kempton, W. (1994). Global environmentalism: A challenge to the postmaterialism thesis? *Social Science Quarterly, 75,* 245–269.

Damon, W. (1977). *The social world of the child.* San Francisco: Jossey-Bass.

Ginsburg, H. P. (1997). *Entering the child's mind: The clinical interview in psychological research and practice.* Cambridge: Cambridge University Press.

Helwig, C. C. (1995). Adolescents' and young adults' conceptions of civil liberties: Freedom of speech and religion. *Child Development, 66,* 152–166.

Hershey, M. R., & Hill, D. B. (1977–78). Is pollution "a White thing?": Racial differences in preadults' attitudes. *Public Opinion Quarterly, 41,* 439–458.

Holland, D. (1997). Selves as cultured: As told by an anthropologist who lacks a soul. In R. D. Ashmore, & L. Jussim (Eds.), *Self and identity* (pp. 160–190). New York: Oxford University Press.

Howe, D., Kahn, P. H., Jr., & Friedman, B. (1996). Along the Rio Negro: Brazilian children's environmental views and values. *Developmental Psychology, 32,* 979–987.

Inglehart, R. (1995). Public support for environmental protection: Objective problems and subjective values in 43 societies. *PS: Political Science & Politics,* 28(1), 57–72.

Kahn, P. H., Jr. (1991). Bounding the controversies: Foundational issues in the study of moral development. *Human Development, 34,* 325–340.

Kahn, P. H., Jr. (1994). Resolving environmental disputes: Litigation, mediation, and the courting of ethical community. *Environmental Values, 3,* 211–228.

Kahn, P. H., Jr. (1997a). Developmental psychology and the biophilia hypothesis: Children's affiliation with nature. *Developmental Review, 17,* 1–61.

Kahn, P. H., Jr. (1997b). Children's moral and ecological reasoning about the Prince William Sound oil spill. *Developmental Psychology, 33,* 1091–1096.

Kahn, P. H., Jr. (1999). *The human relationship with nature: Development and culture.* Cambridge, Mass.: MIT Press.

Kahn, P. H., Jr., & Friedman, B. (1995). Environmental views and values of children in an inner-city Black community. *Child Development, 66,* 1403–1417.

Kahn, P. H., Jr., & Friedman, B. (1998). On nature and environmental education: Black parents speak from the inner city. *Environmental Education Research, 4,* 25–39.

Kahn, P. H., Jr., & Kellert, S. R. (Eds.) (2002). *Children and nature: Psychological, sociocultural, and evolutionary investigations.* Cambridge, Mass.: MIT Press.

Kahn, P. H., Jr., & Lourenço, O. (1999). Reinstating modernity in social science research—or—the status of Bullwinkle in a post-postmodern era. *Human Development, 42,* 92–108.

Kahn, P. H., Jr., & Lourenço, O. (2002). Water, air, fire, and earth: A developmental study in Portugal of environmental moral reasoning. *Environment and Behavior, 34,* 405–430.

Kellert, S. R. (1997). *Kinship to mastery: Biophilia in human evolution and development.* Washington, D.C.: Island Press.

Kellert, S. R., & Wilson, E. O. (Eds.) (1993). *The biophilia hypothesis.* Washington, D.C.: Island Press.

Kempton, W., Boster, J. S., & Hartley, J. A. (1995). *Environmental values in American culture.* Cambridge, Mass.: MIT Press.

Kidd, Q., & Lee, A. (1997). Postmaterialist values and the environment: A critique and reappraisal. *Social Science Quarterly,* 78(1): 1–15.

Killen, M. (1990). Children's evaluations of morality in the context of peer, teacher-child, and familial relations. *Journal of Genetic Psychology, 151,* 395–410.

Kohlberg, L. (1984). *Essays in moral development: Vol. II. The psychology of moral development.* San Francisco: Harper & Row.

Leopold, A. (1970). *A Sand County Almanac.* New York: Ballantine Books. (Original work published 1949)

Lourenço, O. (1990). From cost-perception to gain-construction: Toward a Piagetian explanation of the development of altruism in children. *International Journal of Behavioral Development, 13,* 119–132.

Martinez-Alier, J. (1995). Commentary: The environment as a luxury good or too poor to be green. *Ecological Economics, 13*(1), 1–10.

Maslow, A. H. (1975). *Motivation and personality.* New York: Viking.

McAdams, D. P. (1997). The case for unity in the (post)modern self: A modest proposal. In R. D. Ashmore & L. Jussim (Eds.), *Self and identity* (pp. 46–78). New York: Oxford University Press.

Mohai, P. (1990). Black environmentalism. *Social Science Quarterly, 4,* 744–765.

Myers, G. (1998). *Children and animals: Social development and our connections to other species.* Boulder, Col.: Westview Press.

Myers, G., & Saunders, C. D. (2002). Animals as links toward developing caring relationships with the natural world. In P. H. Kahn, Jr. & S. R. Kellert (Eds.), *Children and nature: Psychological, sociocultural, and evolutionary investigations* (pp. 153–178). Cambridge, Mass.: MIT Press.

Nash, R. (1973). *Wilderness and the American mind.* New Haven, Conn.: Yale University Press.

Nelson, R. (1989). *The island within.* New York: Random House.

Nucci, L. P. (1981). The development of personal concepts: A domain distinct from moral and societal concepts. *Child Development, 52,* 114–121.

Nucci, L. (1996). Morality and the personal sphere of actions. In E. S. Reed, E. Turiel, & T. Brown (Eds.), *Values and knowledge* (pp. 41–60). Mahwah, N.J.: Erlbaum.

Orr, D. W. (1994). *Earth in mind.* Washington, D.C.: Island Press.

Piaget, J. (1960). *The child's conception of the world.* New Jersey: Littlefield, Adams. (Original work published 1929)

Piaget, J. (1969). *The moral judgment of the child.* Glencoe, Ill.: Free Press. (Original work published 1932)

Piaget, J. (1983). Piaget's theory. In W. Kessen (Ed.), P. H. Mussen (series Ed.), *Handbook of child psychology: Vol. 1. History, theory, and methods* (4th ed., pp. 103–128). New York: Wiley.

Pierce, J. C. (1997). The hidden layer of political culture: A comment on "postmaterialist values and the environment: A critique and reappraisal." *Social Science Quarterly, 78*(1), 30–35.

Rosenberg, S. (1997). Multiplicity of selves. In R. D. Ashmore, & L. Jussim (Eds.), *Self and identity* (pp. 23–45). New York: Oxford University Press.

Shepard, P. (1996). *The others: How animals made us human.* Washington, D.C.: Island Press.

Shweder, R. A., & Bourne, E. J. (1984). Does the concept of person vary cross-culturally? In R. A. Shweder & R. A. Levine (Eds.), *Culture theory: Essays on mind, self, and emotion* (pp. 158–199). Cambridge: Cambridge University Press.

Shweder, R. A., Mahapatra, M., & Miller, J. B. (1987). Culture and moral development. In J. Kagan & S. Lamb (Eds.), *The emergence of morality in young children* (pp. 1–82). Chicago: University of Chicago Press.

Smetana, J. G. (1983). Social-cognitive development: Domain distinctions and coordinations. *Developmental Review, 3,* 131–147.

Smetana, J. G. (1995). Morality in context: Abstractions, ambiguities and applications. In R. Vasta (Ed.), *Annals of Child Development* (Vol. 10, pp. 83–130). London: Jessica Kingsley.

Stone, C. D. (1986). Should trees have standing? Toward legal rights for natural objects. In D. VanDeVeer & C. Pierce (Eds.), *People, penguins, and plastic trees* (pp. 83–96). Belmont, Calif: Wadsworth. (Original work published 1974)

Thoits, P. A., & Virshup, L. K. (1997). Me's and we's: Forms and functions of social identities. In R. D. Ashmore & L. Jussim (Eds.), *Self and identity* (pp. 106–133). New York: Oxford University Press.

Turiel, E. (1983). *The development of social knowledge.* Cambridge: Cambridge University Press.

Turiel, E. (1998). Moral development. In N. Eisenberg (Ed.), *Social, emotional, and personality development* (pp. 863–932). Vol. 3 of W. Damon (Ed.), *Handbook of child psychology* (5th ed.). New York: Wiley.

Turiel, E. (2002). *The culture of morality: Social development and social opposition.* Cambridge: Cambridge University Press.

Turiel, E., Killen, M., & Helwig, C. C. (1987). Morality: Its structure, functions and vagaries. In J. Kagan & S. Lamb (Eds.), *The emergence of morality in young children* (pp. 155–244). Chicago: University of Chicago Press.

Wainryb, C. (1993). The application of moral judgments to other cultures: Relativism and universality. *Child Development, 64,* 924–933.

Wainryb, C. (1995). Reasoning about social conflicts in different cultures: Druze and Jewish children in Israel. *Child Development, 66,* 390–401.

Wilson, E. O. (1975). *Sociobiology: The new synthesis.* Cambridge, Mass.: Harvard University Press.

Wilson, E. O. (1984). *Biophilia.* Cambridge, Mass.: Harvard University Press.

7

Children's Environmental Identity: Indicators and Behavioral Impacts

Elisabeth Kals and Heidi Ittner

In this chapter we examine which indicators of environmental identity can explain children's commitment to protect the natural environment. The indicators we focus on embrace emotional as well as cognitive variables. Their ability to predict behavior is compared with that of the more traditional knowledge variable. Our questionnaire studies confirm that even as early as primary school, children have clear concepts about the natural environment, its problems and risks, their own ability to reduce the risks as well as the ability of powerful others, and also who they regard as responsible for reducing these risks. This moral reasoning, together with positively experienced emotions, promotes nature-protective behavior.

An experimental intervention program comparing quantitative and qualitative effects of in-class teaching with the effects of outdoor experiences suggests that the development of an environmental identity in children can be triggered even by short-term interventions. Nevertheless, it is likely to be a long-term process requiring both further experiences and a context that facilitates positive experiences with nature, embedded in moral reasoning about nature.

The data presented here mainly concern bats. In addition to being an endangered species, they constitute an interesting topic through which to analyze children's environmental identity because the typical adult reaction toward bats is fear, dislike, and even disgust.

Environmental Identity

In the past, most children in western societies could enjoy the country-side, play in the fields, interact with animals, and so on. In modern times the facilitation of positive experiences in nature cannot be taken for granted. Children in the industrialized world may live at a great distance from natural settings. They are confronted with pervasive environmental problems such as air pollution, contamination of water and soil, overuse of exhaustible resources, endangerment of species, and lack of green areas or nature reserves. These conditions may make it more difficult for them to affiliate or identify with nature than was previously the case.

Without overromanticizing the pretechnological life in a way that ignores its negative aspects, we may consider the concept of environmental identity as something that has been diminished. What does environmental identity mean? The historical roots of this construct can be traced back to the multifaceted theoretical and empirical work on self and identity (see Filipp, 1988; Hausser, 1995). Marcia (1980), for example, defines identity as "a self-structure—an internal, self-constructed, dynamic organization of drives, abilities, beliefs, and individual history" (p. 159). Owing to its complex character, identity is often replaced by the term *self-concept*, on which research has increased dramatically (see Filipp, 1979, 1988; Frey et al., 1997). Both self-concept and identity have a self-constructed organization, but while the self concept is normally limited to cognitive variables, the broader identity concept also includes emotions (Bosma & Kunnen, 2001; Hausser, 1995; Strayer, 2002).

In the context of nature-protective behavior, we focus on one special facet of identity: environmental identity, which builds upon social relations in the same way as the general identity concept (Bierhoff, 2000; Hausser, 1995). An illustrative example would be membership in a nature conservation group, where environmental identity is evoked and stabilized by interaction with the other members, the so-called ingroup, and by magnifying differences with nonenvironmentalists, representing the outgroup (Eigner & Schmuck, 1998).

The construct of environmental identity is still fuzzy. Taking it literally, environmental identity means identifying with the natural environment and its protection. However, for most people, expressing such a strong affiliation to nature is strange. Therefore, it seems more promising not to measure it in a direct way, but indirectly through its potential main indicators. Such indirect operationalization allows for the possibility that environmental identity is not only important for a small, special minority who, for example, engage actively in fighting pollution, but also for average people who care about the environment without giving it top priority in their lives or inner selves.

In this chapter we examine the developmental aspect of environmental identity from two perspectives. First, we concentrate on the environmental identity of children as an aspect of their sense of self that includes a positive emotional reaction to nature and moral reasoning about it. Second, we suggest ways in which environmental identity emerges and develops.

Although the development of an environmental identity is a lifelong learning process, it has its roots in an early age. It is important to accompany and influence this process from the very beginning in a positive and constructive way, because children of today will have to cope with environmental problems produced today (Pawlik, 1991), which they neither chose nor benefit from (Montada & Kals, 2000). Therefore, it is crucial to understand the development and behavioral impact of children's environmental identity to prepare them for future challenges.

Indicators of Environmental Identity

To select relevant indicators of environmental identity, our starting point was the existing theoretical and empirical research on the prediction of nature-protective behavior. We did not limit our search to one specific theoretical tradition; instead, we combined two traditions, which developed independently, to explain why people act protectively toward nature and accept, for example, personal sacrifices and loss of time or money to ensure this protection. These are (1) models explaining the behavior by emotional affinity toward nature (Kals, Schumacher, & Montada, 1999) and (2) the cognitive theories of moral reasoning (Marcia, 1980), which are applied to nature-protective behavior.

Emotional Affinity toward Nature

Emotional affinity toward nature has proven to be a strong empirical base of nature-protective behavior (Kals, Schumacher, & Montada, 1999). This relationship is based upon the biophilia hypothesis, which claims that humans possess a biologically based attraction to nature and that their well-being depends to a great extent on their relationship with the surrounding natural world (see Eckardt, 1992; Kellert, 1997). Data confirm that direct encounters with nature (e.g., playing outdoors, experiencing nature with all five senses, consciously enjoying the change of seasons) promote affinity toward nature and subsequently identification with it (see Kaplan & Kaplan, 1989; Seel, Sichler, & Fischerlehner, 1993). Five research examples might illustrate this:

• Qualitative analyses (such as content analysis of pupils' essays) conducted by Fischerlehner (1993) support the basic assumption of the positive effects of experiences with nature on environmental attitudes and behaviors.

• Finger (1994) demonstrated that experiences with nature are even more powerful predictors of nature-protective behavior than environmental value orientations.

• Langeheine and Lehmann (1986) have shown that concrete experiences of nature explain a willingness and behavior to protect it. This is especially the case when the effects are reinforced by family norms to treat objects and values carefully.

• Eigner and Schmuck (1998) conducted semistructured interviews with people who engage in proenvironmental behavior within conservation groups. They found that nature experiences played a big role in the childhood of the activists and led to a great appreciation of nature.

• In our own research on adult populations we find that emotional affinity toward nature can be traced back to present and past experiences with the natural environment. Emotional affinity is as powerful in predicting nature-protective behavior as is indignation about insufficient proenvironmental behavior of others and an interest in nature (Kals, Schumacher, & Montada, 1999).

Moral Reasoning about Nature

Moral reasoning about nature involves a shift from the short-term interests of the acting individual to the long-term ecological interests of the whole society. This shift of interests derives from the fact that the benefits of nature-protective behaviors (e.g., in the form of reduced environmental risks) are socialized, whereas their burdens and costs are individualized. The commons dilemma and the more specific term, the *socioecological dilemma*, embody this conflict of interest (Hardin, 1968; Platt, 1973). To overcome the conflict, it is necessary to judge ecological problems and their solutions, not from a self-centered, but from a moral perspective (Kals & Ittner, 2000).

Such a moral perspective includes concerns about ecological problems, the internal and external belief of being able to reduce environmental problems, and the internal and external attribution of responsibility. The more people are aware of ecological problems and the more they recognize efficient strategies and the means to reduce them, the more they are willing to act to protect nature (DeYoung, 1986; Kals, 1996b; Schahn, 1996; Sivek & Hungerford, 1989–90). The same behavioral impact can be found for unselfish, internal attribution of environmental responsibility (Kaiser, Fuhrer, Weber, Ofner, & Bühler-Ilieva, 2001; Kals, 1996a; Montada & Kals, 2000; Schahn, 1996). An internal attribution of ecological responsibility becomes more likely the higher the concern about environment problems and the higher the internal control belief that one has efficient means to reduce them (Kals, 1996a).

The general concern about issues relevant for the inner self as well as subjective explanations, anticipations, and the subjective appraisal of possibilities for action are discussed as crucial components of identity that motivate and regulate individual behavior (Hausser, 1995). Owing to the specific conditions expressed in the term the *socioecological dilemma*, ecological responsibility should be a helpful supplementary indicator of environmental identity.

The theoretical roots of moral reasoning as an indicator of environmental identity can be found in the models of moral development and socialization (Montada, 1995). In these models, moral socialization is understood as the internalization of moral norms, values, and attitudes. This internalization is part of a process of moral development, which is described by Lawrence Kohlberg (1984). His well-known research concerns the development of the arguments supporting normative judgments, which were examined by confronting the subjects with moral dilemmas.

Children's Moral Reasoning

In the case of the commons dilemma the interesting question is at what age are children able to recognize the interests of nature and give priority to them, despite egoistic motives or other moral interests such as economic aims or freedom of choice (Eckensberger, Breit, & Döring, 1999)? It is expected that the higher the moral reasoning, the more ecological responsibility, as part of an environmental identity, is internally attributed.

In the literature concerning the cognitive and emotional indicators of environmental identity, one strong limitation can be found: Model building and empirical research are mainly restricted to adults, whereas environmental research on children remains underdeveloped. Furthermore, the few studies on children's environmental attitudes, behaviors, and identities are mainly restricted to qualitative data. For many research topics, such as analyzing children's mental maps of their natural environment (Treinen, 1997), qualitative approaches are appropriate. Nevertheless, their specific limitations should be compensated by additional quantitative analyses.

Since the importance of environmental research on children is often mentioned (see Gebhard, 1994; Maassen, 1993) and was described earlier, the reasons for this lack of data are presumably practical barriers and measurement problems: From what age onward does it make sense to assess complex environmental variables by standardized scales? How can this be done reliably? How can administrative barriers to studies in schools be overcome and parents' commitment obtained?

A Model of Nature-Protective Commitments

Our empirical research is based on a heuristic model that serves to explain various proenvironmental or nature-protective commitments. These commitments are not restricted to single acts, but cover a broad category of proenvironmental behaviors, such as personal renunciations, political protest against polluting activities of others, public calls for prohibitive laws, and so forth. In several longitudinal studies the commitment variables were validated as proxies of manifest behavioral decisions, which were assessed 2 months later (Montada & Kals, 1998).

The central question of the model is: By what variables can the willingness for continued commitment to protect the natural environment be explained? For the prediction of this commitment criterion, three groups of predictor variables are taken into account. From these predictor variables, emotional affinity and moral reasoning form the indicators of environmental identity, which are contrasted with the traditional knowledge variable (figure 7.1).

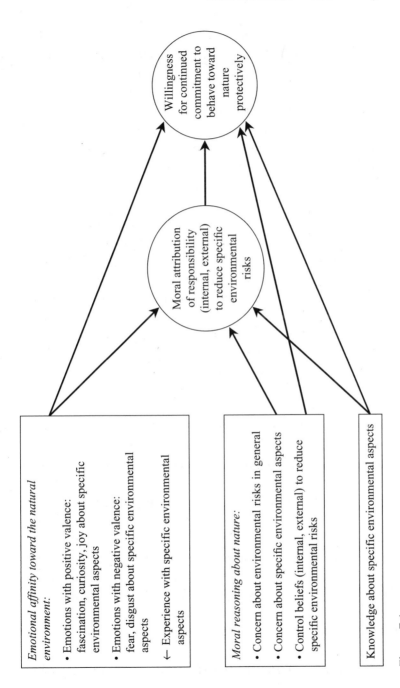

Figure 7.1
A model of nature-protective commitments.

Emotional affinity toward the natural environment, traced back to nature experiences, constitutes the first predictor variable group. It embraces emotions with positive as well as negative valence, because environmental experiences might lead not only to positively but also to negatively experienced emotions (see Maassen, 1993); for example, fear of environmental phenomena such as flooding or confrontations with wild animals, and disgust caused by natural experiences, such as seeing worm farms or animals killing prey.

Moral reasoning about nature as the second predictor variable group includes three categories of cognitions: general and specific concerns about environmental risks, perceived control over these risks, and attributions of responsibility to reduce these risks. The latter two categories are divided into internal and external beliefs, such as the power of megaactors (see Linneweber, 1997). Internal belief refers to acting as a single person (e.g., economizing use of water in one's household) and acting as a member of a social group (e.g., in a conservation group). The latter dimension of an internal control belief reflects on the social interaction with other members of the ingroup, which plays an especially important role in the foundation of an environmental identity.

Finally, the model assumes that nature-protective behavior is also influenced by knowledge about nature and the natural environment. Increased knowledge is the traditional aim of the great majority of environmental education programs (see Calliess & Lob, 1987; Langeheine & Lehmann, 1986). The more concretely the knowledge variable is measured and oriented toward knowledge about means to protect nature, the more powerful it becomes in explaining nature-protective behavior (e.g., Finger, 1994; Graesel, 1999; Kaiser & Fuhrer, 2000). In our model it embraces various aspects of objective endangerment and possibilities for protecting nature. The knowledge variable might promote moral reasoning about nature, but it is not an indicator of environmental identity. Nevertheless, by integrating this traditional variable into the model, it is possible to test the newly construed indicators of environmental identity against the well-analyzed power of the knowledge variable.

Regarding the structure of the model, moral attribution of environmental responsibility is, in line with previous findings (Kals, 1996a), construed as an intermediate criterion that is influenced by environmental concerns and control beliefs. The attribution of internal responsibility should trace back to internal control belief, whereas external responsibility is mainly based on the perception of the power of external agencies (such as government or industry). The additional impact of emotional affinity variables and environmental knowledge on environmental responsibility should be tested explanatorily.

Research Questions

With this model the following three questions are answered empirically:

1. What descriptive findings can be demonstrated concerning the emotional and cognitive indicators of environmental identity within children?
2. What behavioral impact is exerted by the various emotional and cognitive indicators of environmental identity on nature-protective commitments and decisions? Can they also qualify in comparison with the traditional knowledge variable?
3. By what educational strategies can the predictors of nature-protective behavior be promoted? What impact does in-class teaching exert, compared with outdoor experiences with nature, on the knowledge variable as well as on the newly introduced indicators of environmental identity? What are the implications for the conceptualization of intervention programs?

Our heuristic model is not restricted to a specific environmental problem or to specific nature-protective commitments, but for efficient empirical testing this is necessary. In our empirical studies (Kals, Becker, & Rieder, 1999; Lettenbauer, 1999) we applied the model to the overuse and pollution of water as well as to the endangerment of bats, which are currently the most endangered European mammal (Schober & Grimmberger, 1987; Weishaar, 1992).

Methodology

A set of four studies was conducted to answer the descriptive and predictive research questions. In two samples the model was applied to the protection of bats and tested on primary school children ($N_1 = 175$; mean age, 9 years) and younger secondary school children ($N_2 = 137$; mean age, 13 years). In the other two studies the model focused on the protection of water and tested younger secondary school children

(N_3 = 148; mean age, 13 years) as well as older ones (N_4 = 105; mean age, 17 years).

All variables of the model were assessed by several Likert-type items. The younger the children, the easier and shorter the items and the less differentiated the answering scales. The primary school children, for example, answered simply worded items with a 4-point answering scale that reflected (dis)agreement represented by symbolic faces (☹ ☹ ☺ ☺); the older secondary school children got more differentiated items and answering scales. (The appendix gives sample items for the second sample of younger secondary pupils and the protection of bats. A full description of the materials is available from the authors.)

The overall high quality of all measurement instruments (Kals, Becker, & Rieder, 1999; Lettenbauer, 1999) shows that even children at an average age of 9 years are able to give reliable answers concerning complex environmental issues when standardized scales are used.

First Research Question: Emotional and Cognitive Indicators of Children's Environmental Identity

Looking over the descriptive analyses on the environmental topics tested (protection of bats and water pollution) in the various age groups, two overall empirical findings can be summarized:

1. There is a general endorsement of positively valenced emotional affinity toward the natural environment and a lack of expression of emotions with negative valence.
2. A high moral reasoning about nature is expressed.

These general findings are, in the same way as all other findings in this chapter, exemplified by data on bat protection from the second subsample (younger secondary school children; N = 135). It is illustrated by seven variables (table 7.1).

Although a high standard deviation can be found for the central variable of internal responsibility to protect bats, the means of all cognitive variables of moral reasoning are above 3.5 as the midpoint of the scale. The same is true for positively experienced emotional affinity, whereas disgust and fear about bats are denied on average.

These data demonstrate that even children at such a young age have already developed clear and distinctive indicators of environmental identity, including cognitive and emotional elements. They already expe-

Table 7.1
Variables and descriptive results

Variable	Mean	S.D.
Fascination with bats	4.52	1.12
Disgust about bats	1.44	.83
Fear about bats	1.64	.96
Concern about the endangering of bats	5.11	.87
Internal control belief to protect bats efficiently	4.73	.91
Attribution of internal responsibility to protect bats	3.84	1.46
Concern about environmental risks in general	5.49	0.75

Notes: Mean and standard deviation (S.D.); 1 = low agreement to 6 = high agreement; sample: younger secondary school children; $N = 135$).

rience concern about natural risks and problems and have a distinct and varying sense of responsibility to reduce them.

Second Research Question: Behavioral Impact of Indicators of Children's Environmental Identity

To test the behavioral influence of the various indicators of environmental identity, multiple regression analyses were conducted. The commitment variables served as the main criteria, the responsibility appraisals as the intermediate ones. The reliability of the resulting empirical model was successfully proven by excluding different and especially powerful predictor variables.

In the following discussion an exemplifying empirical model is presented that predicts the commitment of younger secondary school children to protect bats by several activities: for example, talking about the endangerment of bats with other people, preventing others from disrupting bats during their winter hibernation, convincing their parents to provide homes for bats, spreading knowledge about their endangerment, and reducing prejudices against them (figure 7.2).

As is shown in figure 7.2, most of the qualifying predictor variables represent cognitive elements of environmental identity. The higher the general concern about environmental risks as well as the specific concern about bats, the higher the external belief that there are efficient measures to protect bats. In addition, the higher the internal attribution of responsibility, the more children are willing to act to protect bats.

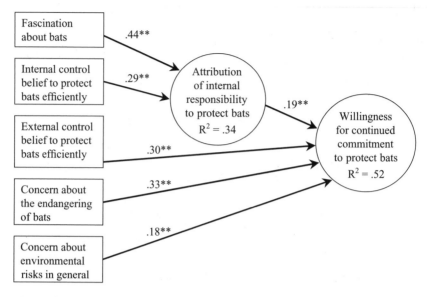

Figure 7.2
Exemplifying empirical model of commitments to protect bats (stepwise multiple regression analysis with standardized regression weights; younger secondary school children; $N = 135$; ** = $p < .01$).

As expected, internal responsibility as the intermediate criterion can be traced back to the internal control belief, but the relationship with fascination about bats shows that it also has an emotional motivation. This fascination is at first glance the only qualifying emotional affinity. However, if this predictor variable is excluded from the predictor set, the remaining two emotions with positive valence qualify for the prediction of internal responsibility instead. These are curiosity about bats and joy observing them in films or in nature. Negatively experienced emotions have no impact on the behavioral or the responsibility criterion. The explained criterion variance of the internal responsibility variable is slightly higher than 30 percent, whereas more than 50 percent of the criterion variance of the commitment variable can be predicted.

This empirical model demonstrates that cognitions of moral reasoning exert a high and stable impact on the prediction of nature-protective commitments. This moral reasoning is not restricted to the specific problem of protecting bats, but is also based on a more general concern about environmental risks. Taking on environmental responsibility is a core variable of the model. It is based upon the belief that one has efficient intervention and action strategies, as well as upon all emotions with positive valence, whereas emotions with negative valence have no impact. That is, fear or disgust about bats does not block the attribution of responsibility to protect bats and to act correspondingly, but has a nonsignificant impact. Only when positive indicators of an identity with the endangered species are developed do children care about their protection.

The nonsignificance of negatively experienced emotions can be only partly explained statistically. Restricted range was not a problem; there was enough variance in the emotional variables. A floor effect is more likely because not many pupils acknowledged the negative items, but at the same time the high level of agreement for positively experienced emotions did not result in nonsignificance. Therefore, the absence of statistical significance for negative emotions is likely to reflect a real insignificance.

In the selected model, the traditional knowledge variable does not qualify for prediction of bat protective behavior, whereas in other samples with different age groups or with water protection as the topic, the impact of our knowledge variable is a little bit higher (Kals, Becker, & Rieder, 1999). However, we consider the small impact of the knowledge variable to result from our way of assessing it. We focused on objective ecological knowledge (e.g., about the way bats live; for an example, see the appendix) instead of more specific knowledge about possibilities to protect bats in the child's own living area. Probably the influence of the latter conception of knowledge would be much higher (see Kaiser & Fuhrer, 2000).

In figure 7.2 as well as in other representative findings, experience in nature has little direct behavioral impact, but strong correlations with emotional affinity toward bats can be found (up to $r = 0.38$, $p < .01$ for joy observing bats). The absence of nature experiences does not trigger negatively valued emotions toward bats, but simply has no effect. All these results can be generalized over different age groups and both environmental problems.

Third Research Question: Promotion of Environmental Identity

The empirical data showed so far that children demonstrate quite stable and differentiated indicators of environmental identity, which have a high impact on nature-protective commitments and decisions. Therefore, the third crucial research question is: How can the various indicators of environmental identity be promoted?

To give preliminary answers to this question, an experimental design was applied within each of the four subsamples. The participating classes were randomized to three conditions:

1. A control group with no intervention;
2. A first experimental group in which environmental issues were taught in the classroom in a traditional way; and
3. A second experimental group in which traditional teaching was supplemented by direct experiences with either bats or natural water resources.

In both experimental groups (traditional teaching as well as supplementary nature experiences) the teaching section took about 8 hours. The same scales, representing the variables of the model (figure 7.1), were measured three times: first, before the decision was made to take part in this longitudinal study (the theoretical model was tested with these data); second, directly after finishing the teaching lessons; and third, as a follow-up measurement 2 months later.

The results indicate that after the teaching lessons, the indicators of environmental identity became more differentiated in the two experimental groups, but not in the control group. We found that in the experimental groups the explained criterion variance of the empirical models became higher when they were conducted with the data set from the second or third time of measurement than with the data gathered before intervention. At the first time of measurement in the four studies on bats and water, the proportion of explained variance for the responsibility and willingness criteria is about 34 to 42 percent (with a maximum of 52 percent for the willingness of younger secondary school children to protect bats; figure 7.2). At the second time of measurement, about 56 to 60 percent of the criterion variance can be explained. This effect remains stable at the third measurement time.

However, there are only a few significant differences in the model variables between the control and the experimental groups. A stable mean effect can be found for the knowledge variable as the traditional variable of school teaching (t_1: $F_{(2;80)} = 0.43$, $p > .05$; t_2: $F_{(2;80)} = 7.14$, $p < .01$; t_3: $F_{(2;80)} = 16.86$, $p < .01$). This is exemplified by the multivariate analysis of variance deriving again from the data set on bats for the younger secondary school children (figure 7.3).

Concerning the indicators of environmental identity, most variables show differences between the experimental and the control groups, but only in some cases were these tendencies (as mean effects) significant. Fascination about bats is selected as a representative result (t_1: $F_{(2;81)}$ = 0.40, $p > .10$; t_2: $F_{(2;81)}$ = 6.68, $p = .10$; t_3: $F_{(2;81)}$ = 9.35, $p = .05$; figure 7.4).

The knowledge variable as well as some indicators of environmental identity increased significantly as a result of the class lectures. Moreover, as expected, additional qualitative data (finishing of sentences as well as teacher ratings) show homogeneously that positively experienced emotional indicators of environmental identity especially were raised during and directly after the short outdoor experience. These results are especially significant because three methodological factors limit the data:

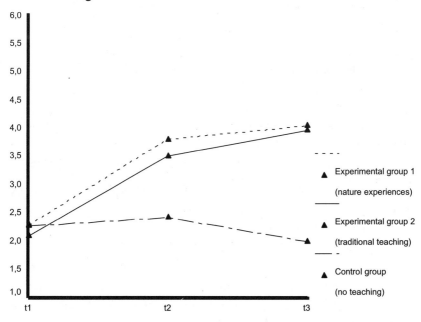

Figure 7.3
Exemplifying experimental effects on the knowledge variable (1 = low knowledge to 6 = high knowledge; younger school children; $N = 85$).

Fascination about bats

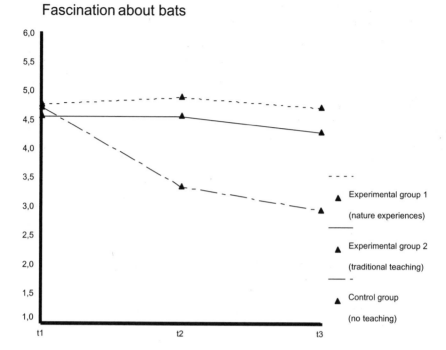

Figure 7.4
Exemplifying experimental effects on the fascination variable (1 = low fascination to 6 = high fascination; younger school children; $N = 85$).

1. The sample sizes in the experimental design were much smaller than those used in testing the theoretical model only at the first time of measurement. Larger samples would be sensitive to even smaller effects.

2. The teaching lessons were restricted to 8 hours and were completed in most cases in 1 or 2 weeks. However, the learning effect should be much deeper when the information is repeated and spread over a longer time.

3. The outdoor experience was part of the 8-hour lecture period and therefore was very short for building emotional bonds with nature.

These limitations can explain why the effects of the experimental treatments were not significant for all variables. For this purpose, larger sample sizes, longer and more continuous teaching, and nature experiences are necessary.

Theoretical and Practical Conclusions

Although the importance of environmental research on children is acknowledged within environmental psychology (see, e.g., Gebhard, 1994), there is little empirical research on it. To address this gap, empirical studies on children's environmental identity and behavior are reported in this chapter. The data confirm that even primary school children clearly express moral reasoning about natural problems. Furthermore, cognitive and emotional indicators of environmental identity exert a stable impact on children's engagements in the protection of nature, whereas negatively experienced emotions are not significant.

Theoretical Model Building

Concerning theoretical model building, the finding that even young children are able to reason morally about nature and have clear emotions toward it strengthens the importance of morality and responsibility in the context of environmental issues. It is possible to overcome commons or socioecological dilemmas by taking on ecological responsibility and identifying with the interests of nature. Children, as well as adolescents (Kals, 1996a), are concerned about the protection of nature, independent of direct personal profits deriving from it. In particular, positive emotions such as fascination, curiosity, and joy play a powerful role in the process of building identity and behavior, whereas negative emotions, such as fear or disgust, are of nearly no significance. This allows a focus upon a positive model of environmental identity, including the behavioral aspects of moral reasoning and emotional affinity.

In future tests of the models, four special aspects should be considered:

1. It would be helpful to include a direct measurement of children's environmental identity. Using this variable, one could examine the extent to which indicators of this identity form a conscious identity upon which children reflect.

2. Furthermore, it would be interesting to examine how the environmental identity changes over time. For this purpose, further longitudinal studies are necessary.

3. The development of an environmental identity should be compared with one competing with nature protection, such as identifying with the aims of promoting economic growth, a high living standard, or a further increase in individual mobility. What, for example, are the differences in

the expression and framing of people's identity when they belong to conservation groups or to motor sport clubs?

4. We need experimental studies on the influences that affect the development of these various identities. Such experimental studies could also address awareness of environmental identity by speaking about and discussing this construct with children. To measure such influences, it is possible and adequate to use standardized questionnaires, even at an early age.

Practical Implications

Without overlooking these future demands, some practical ideas to promote nature-protective behavior can be derived from the data presented here. It makes sense to address the positive emotional and cognitive indicators of the environmental identity of children (for practical examples, see Bilo, Hausen, Schmidt, & Steinkamp, 1991; Peters, 1992). An emotional affinity toward a specific aspect of nature can be promoted by providing positive experiences with nature on a regular basis; for example, by creating outdoor experiences for school classes, which are discussed afterward, that allow children to experience nature through all five senses or by offering other enjoyment of natural surroundings (Gebhard, 1994). It is helpful if these experiences and the subsequent discussion take place with significant others present, because this could stabilize a group identification as a foundation for an environmental identity.

Children not only feel affected by natural problems, they also have concepts about and motivations to reduce them. This motivation is in part based on complex morally relevant reasoning, such as internal attribution of ecological responsibility. Moral reasoning can be promoted by intervention techniques, such as group discussions, contrasting techniques, public statements of social models and so forth, which need to be adapted to the specific age group (see Calliess & Lob, 1987; Langeheine & Lehmann, 1986). Again, the positive effects of group interventions should be used because they should also increase the internal control belief of being able to reduce environmental problems as a member of a group in which others are thinking and acting in the same proenvironmental way. All intervention strategies should be evaluated with regard to their ability to suggest theoretical conclusions and to be improved in the future.

An example of an intervention to promote the proenvironmental engagement of children may illustrate our findings. This example is roughly described for the protection of bats, but its basic components can be applied to many environmental issues. First, children should get some fundamental knowledge about bats and their way of living so that they can discuss their current endangerment. Together they should then think about their own and other people's opportunities to protect bats—but they should also consider reasons why people do not protect bats and even endanger them. Regarding practical experiences outside the classroom, children should have the chance—with little technical support—to see and hear bats flying around (Schober & Grimmberger, 1987). Maybe they could also invite an expert on bats to tell them interesting details or even bring a bat for direct observation. These cognitive- and experience-oriented strategies should awaken and strengthen important positive emotions, such as fascination, joy, or curiosity about these interesting animals. Negative emotions, often based on eerie vampire stories, should be not ignored but should be discussed, so that frightening and disgusting myths lose their power.

Finally, such interventions should not only take place in schools and be directed at children, but should also be directed at adults in many other settings, to foster the building of an environmental identity as a lifelong learning process. Teaching and other intervention forms could be expanded over life through the media, through political discussions of environmental topics, or through involvement with formal (e.g., Greenpeace) or informal (e.g., neighborhood groups taking part in the program Keep America Beautiful) groups engaged in the protection of nature. The stable and sustainable environmental identity that results should help the next generation realize that environmental problems are a communal task with not only moral but also joyful and loving emotional dimensions. To put it another way, as Aldo Leopold said (1949): "We abuse land because we regard it as a commodity belonging to us. When we see land as a community to which we belong, we may begin to use it with love and respect" (p. viii).

Appendix: Examples of Items Regarding Protection of Bats

The sample consists of younger secondary school children. The answering scale ranges from 1 = complete disagreement to 6 = full agreement.

Commitment to protect bats

I am willing to engage in saving the living area and conditions of bats (e.g., meadows with a lot of insects, hedges, and brooks).

Emotions about bats

I am fascinated by bats as they are exceptional animals.
I am curious to learn more about the movement and perception of bats.
I am afraid of bats.
When I think of bats, I feel disgusted.

Control beliefs to protect bats (internal and external)

I am able to protect bats effectively by not disturbing their hibernation.
People who rebuild houses are able to protect bats effectively by not unnecessarily disturbing the living area of bats in old houses.

Attribution of responsibility to protect bats (internal and external)

It is me myself who is responsible for protecting bats.
It is people who rebuild houses who are responsible for the protection of bats.

Concern about the endangering of bats

Bats, which are threatened by extinction, urgently need us to protect them.

Concern about environmental risks in general

If there is no fundamental change, environmental problems will be getting worse and worse within the coming years.

Knowledge about bats

I know a lot of things about bats (e.g., about their sensory organs).

Nature experience with bats

I can well remember having seen a bat.

References

Bierhoff, H.-W. (2000). *Sozialpsychologie* [Social psychology]. Stuttgart: Kohlhammer.

Bilo, M., Hausen, M., Schmidt, R., & Steinkamp, A. (1991). A regional bat-mapping: Suggestion to include schools in work on nature conservation. *Myotis*, 29, 75–82.

Bosma, H. A., & Kunnen, E. S. (Eds.). (2001). Identity and emotion: Development through self-organization. New York: Cambridge University Press.

Calliess, J., & Lob, R. E. (1987). *Handbuch praxis der umwelt- und friedenserziehung* [A handbook: Practice of environmental and peace education]. Düsseldorf: Schwann-Bagel.

DeYoung, R. (1986). Some psychological aspects of recycling: The structure of conservation satisfactions. *Environment and Behavior, 18*, 435–449.

Eckardt, M. H. (1992). Fromm's concept of biophilia. *Journal of the American Academy of Psychoanalysis, 20*(2), 233–240.

Eckensberger, L. H., Breit, H., & Döring, T. (1999). Ethik und barriere in umweltbezogenen entscheidungen: Eine entwicklungspsychologische perspektive [Ethic and barrier in ecologically relevant decisions: A developmental psychological perspective]. In V. Linneweber & E. Kals (Eds.), *Umweltgerechtes Handeln: Barrieren und brücken* [Ecological behavior: Barriers and bridges] (pp. 165–189). Berlin: Springer-Verlag.

Eigner, S., & Schmuck, P. (1998). Biographische interviews mit umwelt- und naturschützern [Biographical interviews with environmentalists]. *Umweltpsychologie, 2*(2), 42–53.

Filipp, S.-H. (Ed.). (1979). *Selbstkonzept-forschung: Probleme, befunde, perspektiven* [Self-concept research: Problems, results, perspectives]. Stuttgart: Klett-Cotta.

Filipp, S.-H. (1988). Das Selbst als Gegenstand psychologischer Forschung [The self as a subject of psychological research]. *Bildung und Erziehung, 41*(3), 281–292.

Finger, M. (1994). From knowledge to action? Exploring the relationships between environmental experiences, learning, and behavior. *Journal of Social Issues, 50,* 141–160.

Fischerlehner, B. (1993). "Die natur ist für die tiere ein lebensraum, und für uns kinder ist es so eine art spielplatz." Über die bedeutung von naturerleben für das 9–13 jährige kind ["Nature is a home for animals and for us children a kind of playground." About the importance of nature experiences for 9–13-year-old children]. In H.-J. Seel, R. Sichler, & B. Fischerlehner (Eds.), *Mensch–natur* [Man–nature] (pp. 148–163). Opladen, Germany: Westdeutscher Verlag.

Frey, D., Krahé, B., Rustemeyer, R., Weber, H., Baltes, P. B., Brandstädter, J., & Filipp, S.-H. (1997). Kommentare zu "Das selbst im lebenslauf—sozialpsychologische und entwicklungspsychologische perspektiven" [Comments on the special issue on social psychological and developmental psychological perspectives on the self]. *Zeitschrift für Sozialpsychologie, 28*(1–2), 129–157.

Gebhard, U. (1994). *Kind und natur* [Child and nature]. Opladen, Germany: Westdeutscher Verlag.

Graesel, C. (1999). Die rolle des wissens beim umwelthandeln—oder: Warum umweltwissen träge ist [The role of knowledge in environmental behavior—or: Why environmental knowledge is inert]. *Unterrichtswissenschaft, 27*(3), 196–212.

Hardin, G. (1968). The tragedy of the commons. *Science, 162,* 1243–1248.

Hausser, K. (1995). *Identitätspsychologie* [Psychology of identity]. Berlin: Springer-Verlag.

Kaiser, F.-G., & Fuhrer, U. (2000). Wissen für ökologisches handeln [Knowledge about environmental behavior]. In H. Mandl & J. Gerstenmaier (Eds.), *Die kluft zwischen wissen und handeln* [The gap between knowledge and behavior] (pp. 51–71). Göttingen: Hogrefe.

Kaiser, F.-G., Fuhrer, U., Weber, O., Ofner, T., & Bühler-Ilieva, E. (2001). Responsibility and ecological behaviour: A meta-analysis of the strength and the extent of a causal link. In A. E. Auhagen, & H.-W. Bierhoff (Eds.), *Responsibility: The many faces of a social phenomenon* (pp. 109–126). London: Routledge.

Kals, E. (1996a). *Verantwortliches umweltverhalten* [Responsible environmental behavior]. Weinheim, Germany: Psychologie Verlags Union.

Kals, E. (1996b). Are proenvironmental commitments motivated by health concerns or by perceived justice? In L. Montada & M. Lerner (Eds.), *Current societal concerns about justice* (pp. 231–258). New York: Plenum.

Kals, E., Becker, R., & Rieder, D. (1999). Förderung umwelt- und naturschützenden handelns bei kindern und jugendlichen [Promotion of proenvironmental behaviors of children and adolescents]. In V. Linneweber & E. Kals (Eds.), *Umweltgerechtes handeln: Barrieren und brücken* [Ecological behavior: Barriers and bridges] (pp. 191–209). Berlin: Springer-Verlag.

Kals, E., & Ittner, H. (2000). Ökologisch relevante lebensqualitäten: Vom singular zum plural [Ecologically relevant qualities of life: From singular to plural]. In M. Bullinger, J. Siegrist, & U. Ravens-Sieberer (Eds.), *Lebensqualitätsforschung* [Research on the quality of life]. *Jahrbuch der medizinischen psychologie,* [Annual review of medical psychology] Vol. 18 (pp. 368–382). Göttingen: Hogrefe.

Kals, E., Schumacher, D., & Montada, L. (1999). Emotional affinity towards nature as a motivational basis to protect nature. *Environment & Behavior, 31*(2), 178–202.

Kaplan, R., & Kaplan, S. (1989). *The experience of nature.* Cambridge: Cambridge University Press.

Kellert, S. R. (1997). *Kinship to mastery: Biophilia in human evolution and development.* Washington, D.C.: Island Press.

Kohlberg, L. (1984). *Psychology of moral development.* San Francisco: Harper: Row.

Langeheine, R., & Lehmann, J. (1986). Die bedeutung der erziehung für das umweltbewusstsein [The importance of education for ecological awareness]. Kiel, Germany: Institut für die Pädagogik der Naturwissenschaften (IPN).

Leopold, A. (1949). *A Sand County almanac and sketches here and there.* New York: Oxford University Press.

Lettenbauer, K. (1999). Umwelterziehung im jugendalter [Environmental education in adolescence]. Unpublished thesis, research area I, psychology. University of Trier, Trier, Germany.

Linneweber, V. (1997). Psychologische und gesellschaftliche dimensionen globaler klimaänderungen [Psychological and social dimensions of global climate change]. In K. H. Erdmann (Ed.), *Internationaler naturschutz* [International protection of nature] (pp. 117–143). Berlin: Springer-Verlag.

Maassen, B. (1993). *Naturerleben mit kindern und jugendlichen* [Nature experiences with children and teenagers]. In H. G. Homfeldt (Ed.), *Erlebnispädagogik* [Pedagogy of experience] (pp. 181–189). Baltmannsweiler, Germany: Schneider-Verlag Hohengehren.

Marcia, J. E. (1980). Identity in adolescence. In J. Abelson (Ed.), *Handbook of adolescent psychology* (pp. 159–187). New York: Wiley.

Montada, L. (1995). Moralische entwicklung und sozialisation [Moral development and socialization]. In R. Oerter & L. Montada (Eds.), *Entwicklungspsychologie* [Developmental psychology] (pp. 862–894). Weinheim, Germany: Psychologie Verlags Union.

Montada, L., & Kals, E. (1998). A theory of "willingness for continued responsible commitment": Research examples from the fields of pollution control and health protection (Berichte aus der arbeitsgruppe "verantwortung, gerechtigkeit, moral" Nr. 114). Research area I, psychology, University of Trier, Trier, Germany.

Montada, L., & Kals, E. (2000). Political implications of environmental psychology. *International Journal of Psychology, 35*(2), 168–176.

Pawlik, K. (1991). The psychology of global environmental change: Some basic data and an agenda for cooperative international research. *International Journal of Psychology, 26,* 547–563.

Peters, G. (1992). Imagepflege für fledermäuse—ein experiment im deutschunterricht [Prestige advertising of bats—an experiment in German class]. *Dendroscopos, 19,* 5–14.

Platt, J. (1973). Social traps. *American Psychologist, 28,* 641–651.

Schahn, J. (1996). *Die erfassung und veränderung des umweltbewusstseins* [Assessment and change of environmental consciousness]. Frankfurt: Lang.

Schober, W., & Grimmberger, E. (1987). *Die fledermäuse Europas: Kennen—bestimmen—schützen* [Bats of Europe: Know—Determine—Protect]. Stuttgart: Frauckh.

Seel, H.-J., Sichler, R., & Fischerlehner, B. (Eds.). (1993). *Mensch—natur* [Man—nature]. Opladen, Germany: Westdeutscher Verlag.

Sivek, D. J., & Hungerford, H. (1989–90). Predictors of responsible behavior in members of three Wisconsin conservation organizations. *Journal of Environmental Education, 21,* 35–40.

Strayer, J. (2002). The dynamics of emotions and life cycle identity. *Identity, 2*(1), 47–79.

Treinen, C. (1997). Mental maps—am beispiel der umweltwahrnehmung von kindern. Eine empirische untersuchung [Mental maps—for example of the environmental perception of children]. Unpublished thesis, research area VI, geoscience. University of Trier, Trier, Germany.

Weishaar, M. (1992). Landschaftsstrukturen, unersetzliche elemente im fledermausschutz [Structures of landscape, irreplaceable elements in protection of bats]. *Dendroscopos, 19,* 15–18.

II

Experiencing Nature in Social and Community Contexts

8

The Human Self and the Animal Other: Exploring Borderland Identities

Linda Kalof

We polish an animal mirror to look for ourselves.
—Donna J. Haraway, *Simians, Cyborgs, and Women*

It is largely uncontested that human identity and the human self are constructed in social relationships within a specific historical and cultural context. Less widely accepted in Western culture, however, is that social relationships consist of both humans and nonhumans and that human identity is developed through relationships with nonhuman others. Because of the human penchant for conceptualizing self–other, same–different and culture–nature as binary opposites, animals are not considered part of the human identity equation. This dualistic thinking fosters a hyperseparation in which differences between humans and animals, or culture and nature, are emphasized and magnified (Plumwood, 1993). Indeed, such a separation and difference even provides a critical marker for identity development in the argument that all socially constructed identities depend on their opposites for meaning (Anderson, 1998; Shepard, 1996).

Scholars have recently interrogated the rigid nature of the traditional boundaries drawn between the human and the nonhuman (Haraway, 1989, 1991, 1997; Michel, 1998; Plumwood, 1993). As Emel and Wolch (1998) note, feminism and postmodernism have done irreparable damage to the ideology that there is an impassable gap between humans and animals, reminding us that these boundaries were built by intellectuals in the first place. Whether the discourse is framed by similarity, difference or blurred boundaries, animals are critical contributors to the construction of human identity and a "relational self." A relational self is grounded in friendship, solidarity, caring, and the recognition of both kinship with and difference from others (Plumwood, 1993). To what degree do humans perceive or recognize a similarity or an affinity with other animals? In this chapter I explore this question using data from college students and their evaluations of nonhuman animals.

Background

Similarity and Dissimilarity

Attraction toward similar others is often explained using the "similarity principle" (Plous, 1993). We give more consideration to those who are perceived as similar to us—we more frequently offer help to similar others; we are more concerned about the pain of similar others; and we punish more harshly when a victim is similar to us. Animals are largely perceived as dissimilar (inferior, unattractive, homogeneous, and more tolerant of pain), a perception that justifies human maltreatment by construing animals as an outgroup (Plous, 1993). Indeed, it has been argued that the most important factor in determining our attitudes toward animals is how similar to us we think they are (Lawrence, 1995).

It comes as no surprise that dislike of dissimilar, distanced others is closely linked to racism and sexism, in which marked categories of people metaphorically resemble animals (Ritvo, 1995). Our sorry tradition of oppressing others is replete with examples where similarities are drawn between humans considered inferior and nonhuman animals (Adams, 1990, 1994; Emel & Wolch, 1998; Gould, 1996; Spiegel, 1996; Shipman, 1994). For example, Desmond (1999) notes that the display of human "freaks" in cages shows the "disenfranchisement of the physically different and their banishment across the species lines to the objectified status" of animals (p. 157).

Borderland Identities

In a revolutionary shift away from hyperseparation—exclusion and dislike based on difference—some scholars have begun to advocate "borderland identities." In the borderlands, boundaries become blurred and new understandings that recognize both similarities and differences emerge. The awareness of blurred boundaries between human and natural entities provides a framework for seeing identities as products of interrelations between the human and nonhuman (Michel, 1998).

Dualistic categories are now conceptualized as closely related to social practice and are highly variable in their construction and deployment (Mullin, 1999). Donna Haraway has written extensively (1989, 1991, 1997) on how human identity is defined in relationships with nonhuman others in historical and political contexts. She argues for a new concept of identity and social relations—associations of affinity created from multiple interactions among people, objects, and animals.

Affinity, Kinship, and the Relational Self

Affinity relationships are based on choice, not blood (Haraway, 1991), emphasizing the formation of coalitions usually centered around a political agenda or a common concern. These relationships recognize both kinship with and difference from others, a prerequisite for dismantling self–other dualisms and building a relational self with a noninstrumental relationship with nature (Plumwood, 1993). The relational self is interested in "the flourishing of earth others and the earth community among its own primary ends, and hence respects or cares for these others for their own sake" (Plumwood, 1993, p. 154). Thus, "similarity," "difference," and "identity" all take on new meanings in the postmodern feminist context of blurred boundaries between nature and culture. Identity is considered fluid—a practice, not a category, and a distinctive move away from essentialist conceptual categories such as gender and race. Identities are actively constructed performances rather than preexisting roles (Bucholtz, Liang, & Sutton, 1999). And it is the "contradictory, partial, and strategic" aspects of identity that signify the importance of thinking about *affinity* in place of *identity* (Haraway, 1991, p. 155). Thus the emphasis is placed on our kinship with those with whom we have a community of interest rather than a sameness of essential character.

Empirical Evidence
Some recent research on gender, race, and concern for nonhuman animals provides good examples of the contradictory, partial, and fluid nature of identity that morphs into affinity—a perceived resemblance to those with whom one has a common interest. For example, based on survey data from American adults, Peek, Dunham, and Dietz (1997) have argued that women's high levels of support for animal rights are not

explained by an ethic of care (Gilligan, 1982), but rather by women's inferior position in the social structure, an experience of oppression that gives rise to empathy for other oppressed groups. Elsewhere I have documented the shifting nature of identity markers (such as gender, race, and age) usually associated with high or low levels of concern for nonhuman species (Kalof, 2000). Using an analysis that identified groups of respondents who had similar attitudes about animal concern, I found a variety of multilayered discourses of concern for other animals; some discourses supported traditional ideas of human dominion over nature, some opposed those ideas. But all discourses were multiple, overlapping maps of social characteristics, providing evidence of the fluid and shifting nature of identity markers. For example, Hispanic and white women with high altruism had similar "positive" attitudes about animals, but young nonwhite women with high altruism had similar "negative" attitudes about animals. Thus, while altruism and gender were stable markers of both discourses, ethnicity was an unstable and contradictory characteristic.

In spite of the presence of the similarity–difference and kin–not kin problematics in our discourse about animals, few researchers have undertaken empirical studies of human–animal affinity relations. Most of the studies that have focused on human identification with animals come primarily from anthropology and psychology. For example, Geertz (1975) studied the use of animals to display signs and signification of human identity. In his analysis of cockfighting in Bali, Geertz observed that the cockfight is central to Balinese life and is the means by which the Balinese portray themselves to themselves. Balinese men have a very close attachment to their cocks (the same pun applies in Balinese as in English), and because the Balinese have a strong fear and hatred for anything animal-like, the men's identification with the birds is both an exaggeration of the male ego and a connection to the feared (Geertz, 1975). (It is interesting to note that Geertz's metasocial commentary on Bali was recently criticized by a contradictory analysis that called into question his lack of concern with history and diverse viewpoints; see the discussion in Mullin, 1999, p. 211.)

Another study of human–animal affinity (or similarity–difference) was conducted by Opotow (1993). She used an experimental design to examine the conditions under which high school students extended justice and concern to a nonhuman animal based on the animal's perceived similarity or difference in relation to humans. The animal's similarity to humans (the same biological needs and social behavior, rearing of the young, protection of territory) had no influence on the students' willingness to provide the animal with resources, fairness, a personal sacrifice, or protective measures. However, the students were willing to extend substantial justice and concern to animals perceived as useful to humans. Opotow (1993) argued that perceiving another as similar is not enough to create personal self-interest in the animal's well-being. Indeed, students *decreased* their extension of justice to the animal when it was perceived as both similar to them and in a high conflict situation with humans (for example, the human need for the animal's habitat). Thus, context is crucial for the extension of the scope of justice and for what kinds of utility and similarity matter to people (Opotow, 1993). In a somewhat similar study, college undergraduates were more likely to attribute cognitive states (ability to trick another animal, ability to recognize self in a mirror, ability to determine the difference between being tripped over or being kicked) to animals perceived as similar to humans and to animals higher on the phylogenetic scale (Eddy, Gallup, & Povenelli, 1993).

Researchers have also studied human attribution of meaning and feelings of attachment to nonhuman animals. For example, the connotative meanings of twelve animal names (such as "alligator") were examined by Baenninger, Dengelmaier, Navarrete, and Sezov (2000). Using Osgood, Suci, and Tannenbaum's (1957) semantic differential technique, they found that college students identified two groups of animals: carnivorous predators and pets. The authors suggested that the groupings represented "scary" and "cute" dimensions of meaning. The response patterns were not influenced by any of the known demographic characteristics of the sample (gender, marital status, urban-rural-suburban residence, age, zoo attendance, watching wildlife or animal programs on television). The companionship of pets also emerged as a significant predictor of empathy with animals in Paul (2000).

Semantic differential scales were also used by Finlay, James, and Maple (1988) in their study of people's perceptions of animals in different contextual settings (the caged zoo, the naturalistic zoo, and the wild). They reported that as expected, the animals in the caged zoo were rated less favorably than animals in the naturalistic zoo or in the wild and, while zoo animals were perceived as passive and restricted, animals in the wild were seen as active and free.

One study of particular importance to my work here was Brown's (1996) exploration of college students' feelings of attachment to nonhuman animals. He found that the students' subjective ties to animals fell into two categories: understanding and closeness. The understanding dimension (the extent to which one is capable of understanding the emotions of particular animals) embraced "spatial and evolutionary proximity" (1996, p. 11). The dimension of closeness was applied to animals with whom students had "feelings of emotional closeness and (ostensibly) physical similarity" (p. 13). Brown concluded that the understanding dimension represented an "objective" engagement with animals based on either physical similarity (chimpanzee, gorilla) or intimate association (dog, cat, horse). The closeness dimension represented a "social" tie, such as feelings of emotional attachment to the idea that one has social characteristics in common with animals (such as the perseverance of the turtle or the loneliness of the wolf) or identification with fashion adornments (ladybug).

In summary, while there is much theory to foreground the construction of human identity and a sense of self in affinity relationships with nonhuman species, precious few empirical attempts have been made to flesh out this important process. And, while studies have addressed human attributions of meaning, value, and attachment to other animals, I am not aware of any research that attempts to find an association between one's description of self and one's description of nonhuman species, a good test of affinity or viewing oneself as kin to the other.

Analysis of Perceived Characteristics of Self and Other

To explore human perceptions of affinity with or hyperseparation from a nonhuman animal stimulus, I examined (1) college students' evaluations of a videotaped nonhuman animal image using semantic differential bipolar adjective scales (Osgood, Suci, and Tannenbaum, 1957), (2) the students' evaluations of themselves along the same bipolar dimensions, and (3) the association between the animal image evaluations and the self evaluations (with negative associations indicating hyperseparation or perceived difference and positive associations indicating affinity or perceived similarity with the animal).

My sample consisted of 180 students, 33.9 percent of whom were men and 66.1 percent women, most between 17 and 25 years of age. There was a wide diversity of race and ethnic categories in the sample: 12.8 percent were Black or African American; 55.6 percent White or Caucasian; 9.4 percent Hispanic, Chicano, or Spanish-speaking American; 12.8 percent Asian, Pacific Islander, or Filipino; and 9.4 percent were American Indian, East Indian, Middle Eastern, or biracial.

The students used a 6-point Likert scale to indicate their degree of agreement with twenty-six attitude statements on the meaning and value of nonhuman species and then watched two short videotapes of television advertisements portraying the instrumental use of animals. In one advertisement, a man showed his love for cheese by going on a date with and kissing a cow (the cow was used for making cheese). In the second advertisement, a mother and children were shown first in a dirty medieval market, then in a turn-of-the century market, surrounded by live (but almost-slaughtered) and already-slaughtered animals. The ad closed with a short view of the family in a modern, clean supermarket made possible by plastics (the almost-slaughtered and already-slaughtered animals were used for food). Students were then asked to think of the one animal in the ads that stood out most in their minds and to evaluate the animal along a 1 to 6 dimension between bipolar adjectives (described in the following section). Finally, the students were asked to evaluate themselves using the same bipolar adjectives they used to describe the animal image.

Dimensions of the Animal Image

The twenty-one bipolar adjective pairs were generated from a preliminary reading of a sample of randomly selected advertisements that portrayed nonhuman animals used as tools (Lerner & Kalof, 1999). The bipolar pairs were bad/good, timid/bold, kind/cruel, safe/dangerous, vulnerable/not vulnerable, strong/weak, beautiful/ugly, exploited/not exploited, tender/tough, foolish/wise, active/passive, valuable/not valuable, dependent/independent, competitive/not competitive, complicated/uncomplicated, dominant/submissive, family/not family, fast/slow, natural/not natural, useful/not useful, and bothersome/not bothersome.

A principal components factor analysis with varimax rotation revealed three distinct dimensions of the animal image. Animal image 1 consisted of the following adjectives (in order of importance; factor loadings less than 0.30 were disregarded): *competitive, fast, complicated, active, dominant, beautiful, family.* Animal image 2 consisted of *kind, good, valuable, safe, not bothersome, useful, natural.* Animal image 3 consisted of *vulnerable, weak, timid, tender, foolish, dependent, exploited.*

Dimensions of the Self Image

The same analytic strategy was used for the self ratings on the twenty-one bipolar adjective pairs. The analysis revealed three distinct dimensions of the self image. Self image 1 consisted of the following adjective ratings of self (again, presented in order of importance): *active, dominant, bold, strong, fast, competitive, beautiful, complicated.* Self image 2 consisted of *good, kind, safe, not bothersome, useful, not exploited, family, valuable.* Finally, self image 3 consisted of *vulnerable, tender, foolish, natural, dependent.*

Three factors were extracted from the factor analysis of the animal-image ratings, and three factors were extracted from the factor analysis of the self-image ratings. The three self-image factors and the three animal-image factors were remarkably similar. Thus for both self and animal ratings, image 1 was labeled *agency*; image 2, *benevolence*; and image 3, *fragility.* To show the similarity of the ratings, table 8.1 provides the adjectives that comprised each self and animal dimension.

Dimensions of Attitudes toward Nonhuman Species

Three distinct factors were extracted from a factor analysis (principal components, varimax rotation) of the responses on the twenty-six statements concerning the meaning and value of nonhuman species. I labeled the factors *ecocentric attitudes* (centered by ecosystems in which entire species and ecosystems are morally important in their own right and intrinsically valuable, not just individual animals); *sanguinolent attitudes* (centered by bloodshed, where hunting and killing individual animals for human recreation or use is morally acceptable) (*Webster's Third New International Dictionary* defines sanguineous as "of, relating to, or involving bloodshed." I chose this label over dominionistic attitudes or some variation of dominance or power because of the overwhelming

Table 8.1
Dimensions of animal image and self image[a]

Animal image Dimension 1 (agency)	Animal image Dimension 2 (benevolence)	Animal image Dimension 3 (fragility)
Competitive	Kind	Vulnerable
Fast	Good	Weak
Complicated	Valuable	Timid
Active	Safe	Tender
Dominant	Not bothersome	Foolish
Beautiful	Useful	Dependent
Family	Natural	Exploited

Self image Dimension 1 (agency)	Self image Dimension 2 (benevolence)	Self image Dimension 3 (fragility)
Active	Good	Vulnerable
Dominant	Kind	Tender
Bold	Safe	Foolish
Strong	Not bothersome	Natural
Fast	Useful	Dependent
Competitive	Not exploited	
Beautiful	Family	
Complicated	Valuable	

[a] From principal components factor analysis, varimax rotation, loadings greater than 0.30.

presence of statements dealing with killing individual animals that loaded on the factor.); and *humanocentric attitudes* (centered by concern for human welfare that supersedes consideration of both nonhuman species and individual animals). See the appendix for a list of the statements that comprised each dimension of concern for nonhuman species.

Hyperseparation or Affinity?

Table 8.2 shows that a benevolent rating of self had a substantial positive correlation with rating the animal as benevolent, and, while it was not statistically significant at conventional levels, there was a positive correlation between the agentic self and the agentic animal. There was no relationship between a fragile rating of self and rating the animal as fragile. Since the associations were positive, the perception was one of similarity or affinity with the animal, not hyperseparation.

Ecocentric attitudes were negatively related to a fragile animal image and the benevolent animal image. Sanguinolent attitudes were positively correlated with the agentic animal image. Finally, there was some evidence of a positive relationship between humanocentric attitudes and the fragile animal image, but the association was not significant at conventional levels.

To examine the association between ratings of self and ratings of animals by gender and race, I ran separate bivariate correlations, one each for men, women, white (majority) students, and nonwhite (minority) students. As in the overall correlational analyses, only one self rating was correlated with an animal rating in the subgroups: benevolence. The benevolent self was positively related to the benevolent animal, and that pattern of similarity held for all subgroups (women, men, white majority and nonwhite minority).

Finally, a multivariate analysis of variance of the three animal ratings established that one dimension of self (benevolence) had a significant overall effect on the animal ratings, as did ecocentric attitudes, sanguinolent attitudes, and race. Thus, while race was associated with the animal ratings, gender had no overall effect, and there was no gender–race interaction. The univariate analysis showed that most of the variation in the animal ratings was accounted for by: (1) the large positive relationship between the benevolent self rating and the rating of the animal as benevolent, (2) the negative relationship between ecocentric attitudes and rating the animal as fragile, (3) the positive relationship between sanguinolent attitudes and rating the animal as agentic, and (4) the influence of race in rating the animal as fragile (American Indian, East Indian, Middle Eastern, and biracial students rated the animal as very fragile; Asian students rated the animal as not at all fragile; and White, Black, and Hispanic students fell in the middle of these extremes).

Discussion

Using a set of twenty-one bipolar adjective pairs, I found that the college students provided remarkably similar evaluations of themselves and of

Table 8.2
Bivariate correlations

	Agentic self image	Benevolent self image	Fragile self image	Agentic animal image	Benevolent animal image	Fragile animal image	Ecocentric attitudes	Sanguinolent attitudes	Humanocentric attitudes
Agentic self image	1.00	0.00	0.00	0.13*	0.00	-0.07	0.04	-0.10	-0.06
Benevolent self image		1.00	0.00	0.04	0.30***	-0.13	-0.06	0.01	0.00
Fragile self image			1.00	0.09	0.05	0.02	-0.18**	0.24***	0.05
Agentic animal image				1.00	0.00	0.00	-0.06	0.20***	0.11
Benevolent animal image					1.00	0.00	-0.16**	-0.01	0.11
Fragile animal image						1.00	-0.15**	0.10	0.13*
Ecocentric attitudes							1.00	0.00	0.00
Sanguinolent attitudes								1.00	0.00
Humanocentric attitudes									1.00

*** $p < .01$, ** $p < .05$, * $p < .10$.

a nonhuman animal. The factor structure for the student self evaluations and the evaluations of an animal stimulus were nearly identical. There was no evidence of hyperseparation in which the students emphasized or magnified nonhuman animal differences from their human selves. The most important dimension of similarity was benevolence. A benevolent self (kind, good, safe) consistently rated the animal as benevolent. Also, viewing oneself as agentic (active, fast, competitive) increased the tendency to regard the animal as agentic. The third self dimension, fragility (vulnerable, weak, timid), had only a weak (but positive) association with rating the animal as fragile.

Self-animal rating correlational analyses by gender and race-ethnicity showed no obvious gender or race-ethnic differences in the bivariate relationships. Only one self-animal rating was correlated in the subgroup analyses: the substantial positive association between a benevolent self and the rating of the animal as benevolent, a pattern that held for all subgroups (women, men, white (majority) students and nonwhite (minority) students. It is interesting to note that the bivariate analyses indicated no support for the theoretical argument that women and minorities would show more affinity and less hyperseparation from the animal other than men and the majority students.

Attitudes toward nonhuman species were also related to the animal image ratings. For example, students with sanguinolent attitudes perceived the animal image as agentic. I was not surprised to see a relationship between the sanguinolent "prohunting, killing is acceptable" attitudes and perceiving the animal as fast, competitive, and active. It is also worth noting that the agentic animal dimension included the adjectives *dominant*, *beautiful*, and *family*. The dominance of the animal is a key issue in the hunt, with the kill establishing the dominion of the hunter over the hunted (or culture over nature). Jose Ortega y Gasset (1972) defined the human and nonhuman hunt as "what an animal does to take possession, dead or alive, of some other being that belongs to a species basically inferior to its own" (quoted in Luke, 1998, p. 629). The hunted animal is often described as a beautiful and worthy opponent, particularly if the hunt is a "good" one. It is also common to portray hunting as a temporary union with the natural world (Cartmill, 1995), implying a familial connection between humans and animals, as in Gary Snyder's (1969) argument that a hunter can become physically and psychically one with the animal. However, Kheel (1995) has noted that while the ethic of the hunt might be called *"bio*centric," it is better described as *"necro*centric" since "it is death, not life, that (connects) the hunter with other living beings (a)nd it is death, not life, that elicits feelings of reverence and respect" (p. 107).

The finding that the students with ecocentric attitudes perceived the animal as less fragile and less benevolent may at first blush appear counterintuitive. Perhaps this was the result of the ecocentrists' attempts to deanthropomorphize the animal image because of the belief that ecosystems and entire species are morally important and intrinsically valuable in their own right. Thus a nonhuman animal cannot be described using human characterizations and adjectives such as *bold, strong, good,* and so on. Indeed, ecocentric attitudes were negatively related to all three animal images (and two of the self images).

In an attempt to explain the variation in animal image ratings, I found (again) that the benevolent self was a powerful predictor of the animal ratings, primarily in its positive association with the benevolent animal image. The important presence of benevolence in this study corroborates earlier research that documented the pivotal role of altruism in producing consistent orientations toward nonhuman species among a diverse group of respondents (Kalof, 2000). When the respondents differed in altruistic value orientations, the discourses on concern for animals became complex, conflicting, and contradictory. There is another possible explanation for the emergence of benevolence as the most compelling factor in this study of ratings of self and ratings of the animal other. Benevolence could be considered the most salient characteristic of a relational self.

While it was not important in the bivariate analysis, race was an important predictor of the animal ratings in the multivariate analysis. The student evaluations of the animal as fragile were substantially different based on their race or ethnic group. Asian students perceived the animal as least fragile, and students of other ethnic groups (such as American Indian, East Indian, Middle Eastern) perceived the animal as most fragile. This different reading may be a cultural difference captured by race or ethnic identity, with Asians the least familiar or comfortable with nonhuman animals. For example, in an earlier study (Kalof, 2000), young Asian and black women from U.S. cities and suburbs were similar in their tendency to rank animals on how well they fit into human society, with "good" animals accepting their place in society (pets) and "bad" animals disrupting human life (wildlife that disturbs urban areas). This system of ranking according to how well individuals fit into society and play their expected roles constitutes a sociozoologic system, with an animal's worth and position determined by the individual animal's willingness to accept subordination (Arluke & Sanders, 1996). For example, as Haraway notes, wild, farmed foxes "object to their captivity, including their slaughter" (Haraway 2001, p. 29). The foxes then might be considered bad animals, differentiated from the domesticated-for-food animals such as sheep and cattle—the good animals.

Gender had no overall effect on the animal ratings, nor was there any interaction between gender and race. Thus, in addition to finding no support for the argument that women and minorities would show more affiliation with animals than men and majority students (discussed earlier), gender had no observable influence on perceptions of animals, either alone or in conjunction with race. I believe that this is further evidence of the fluid nature of identity markers, particularly gender, that do not conform to traditional binary concepts such as male–female in the evaluation of attitudes, perceptions, and concern for others. Furthermore, the weak role of gender in explaining variance has been documented in other research on concern for animals. For example, Herzog, Betchart, and Pittman (1991) found gender and sex role orientation together only explained 10 percent of the variance in animal welfare attitudes, and in Driscoll (1992) gender, religion, and pet ownership together explained only 5 percent of the variance in attitudes about the use of animals in a variety of contexts, such as medical research, behavioral research, and product testing. Thus it appears that human evaluations of animals (their use, their welfare, their image) take shape and form in multiple sites, such as gender, race, and nationality, with each site contributing a partial interpretation or a partial reading of the animal.

Blurred boundaries and fluid identities invoke the "destabilized" human self. This destabilized self contains identities both distinct from and shared with natural entities and is constructed in terms of mutuality with nature and other organisms (Michel, 1998). This new conception of identity, this relational self, might help us find better ways of living within ecosystems, ways that respectfully acknowledge our continuity with and our differences from other inhabitants (Hawkins, 1998). I found that the most resilient marker in the evaluation of the nonhuman animal was the evaluation of self, and my human respondents had a tendency to describe themselves and nonhuman animals in similar ways. It is my hope that this pattern portends a kinship identification with the nonhuman other, "as a member of a species and a zoological order" (Haraway, 2000, p. 399). Finally, this emphasis on affinity and kinship avoids essentialism and nominalism (McLaughlin, 1998), both of which are rooted in the master story of Western culture, "a story of conquest and control, of capture and use, of destruction and incorporation" (Plumwood, 1993, p. 196). When perceived as kin, animals provide us with a bond to the natural world (Mason, 1998), and acknowledgment of this bond, this affinity with animals, may in turn provide the inspiration for new, less destructive forms of shared survival.

Appendix: Three Dimensions of Attitudes toward Nonhuman Species

The respondents rated each item on a 1–6 scale, with higher ratings indicating more agreement with the statement.

Ecocentric *(alpha = 0.69)*

1. Zoos should provide more natural conditions for animals, even if that means much higher entrance fees.
2. I would rather pay a higher price for tuna fish than see the tuna industry kill porpoises in their nets.
3. Justice is not only for humans; animals and plants also deserve justice.
4. If any species has to become extinct as a result of human activities, it should be humans.
5. Plants and animals are here to serve humans, they don't have any rights in themselves (reverse scored).
6. Species of animals and plants have intrinsic value, even if not of any use to humans.
7. I would rather see a few humans suffer or be killed than see human damage to the environment cause an entire species to go extinct.
8. Continued existence of wilderness and wildlife is critical to the spiritual well-being of humanity.

Sanguinolent *(alpha = 0.70)*

1. There is nothing wrong with killing individual animals, as in hunting, as long as you don't kill so many the population is threatened.
2. It is all right to kill whales for a useful product as long as the animals are not threatened with extinction.
3. I admire a person who works hard to shoot a big trophy animal like a 600-pound bear.

4. It is all right to kill an animal to make a fur coat as long as the species is not endangered.
5. I see nothing wrong with using steel traps to capture wild animals.
6. Since Eskimos have always hunted the bowhead whale for its meat, they should be allowed to continue even though this whale may be endangered.
7. We must use pesticides harmful to animals and plants if they are needed to maintain our food production.
8. We should become vegetarians to reduce our environmental impact (reverse scored).

Humanocentric *(alpha = 0.68)*

1. Natural resources must be developed, even if the loss of wilderness results in much smaller wildlife populations.
2. I dislike most beetles and spiders.
3. I approve of building on marshes that nonendangered wildlife use if the marshes are needed for housing development.
4. If oil were discovered in Yellowstone Park, it would have to be developed even if it meant harm to the park's plants and animals.
5. I know little about ecosystems or the population dynamics of wild animals.
6. I think rats and cockroaches should be eliminated.
7. Indians have no greater right to hunt wildlife than others no matter what has been granted in past treaties.

References

Adams, C. J. (1990). *The sexual politics of meat: A feminist-vegetarian critical theory.* New York: Continuum.

Adams, C. J. (1994). *Neither man nor beast: Feminism and the defense of animals.* New York: Continuum.

Anderson, K. (1998). Animal domestication in geographic perspective. *Society and Animals, 6,* 119–135.

Arluke, A., & Sanders, C. R. (1996). *Regarding animals.* Philadelphia: Temple University Press.

Baenninger, R., Dengelmaier, R., Navarrete, J., & Sezov, D. (2000). What's in a name: Uncovering the connotative meanings of animal names. *Anthrozoos, 13,* 113–117.

Brown, S. R. (1996). Contributions to the study of animals and society. Paper presented at the International Society for the Scientific Study of Subjectivity.

Bucholtz, M., Liang, A. C., & Sutton, L. A. (Eds.). (1999). *Reinventing identities: The gendered self in discourse.* New York: Oxford University Press.

Cartmill, M. (1995). Hunting and humanity in western thought. *Social Research, 62,* 773–786.

Desmond, J. (1999). *Staging tourism: Bodies on display from Waikiki to Sea World.* Chicago: University of Chicago Press.

Driscoll, J. W. (1992). Attitudes towards animal use. *Anthrozoos, 5,* 32–39.

Eddy, T. J., Gallup, G. G., Jr., & Povinelli, D. J. (1993). Attribution of cognitive states to animals: Anthropomorphism in comparative perspective. *Journal of Social Issues, 49,* 87–101.

Emel, J., & Wolch, J. (1998). Witnessing the animal moment. In J. Wolch & J. Emel (Eds.), *Animal geographies: Place, politics and identity in the nature-culture borderlands* (pp. 1–24). New York: Verso.

Finlay, T., James, L. R., & Maple, T. L. (1988). People's perceptions of animals: The influence of zoo environment. *Environment and Behavior, 20,* 508–528.

Geertz, C. (1975). Deep play: Notes on the Balinese cockfight. In C. Geertz, *The interpretation of cultures* (pp. 412–453). New York: Basic Books.

Gilligan, C. (1982). *In a different voice: Psychological theory and women's development.* Cambridge, Mass.: Harvard University Press.

Gould, S. J. (1996). *The mismeasure of man* (2nd ed.). New York: Norton.

Haraway, D. (1989). *Primate visions: Gender, race, and nature in the world of modern science.* New York: Routledge.

Haraway, D. J. (1991). *Simians, cyborgs, and women: The reinvention of nature.* New York: Routledge.

Haraway, D. J. (1997). *Modest_Witness@Second_Millennium.FemaleMan ©_Meets_OncoMouse™.* New York: Routledge.

Haraway, D. (2000). Morphing in the order: Flexible strategies, feminist science studies, and primate revisions. In S. C. Strum & L. M. Fedigan (Eds.), *Primate encounters: Models of science, gender, and society* (pp. 398–420). Chicago: University of Chicago Press.

Haraway, D. (2001). For the love of a good dog: Webs of action in the world of dog genetics. Invited presentation, the University of Michigan.

Hawkins, R. Z. (1998). Ecofeminism and nonhumans: Continuity, difference, dualism, and domination. *Hypatia, 13,* 158–197.

Herzog, H. A., Jr., Betchart, N. S., & Pittman, R. B. (1991). Gender, sex role orientation and attitudes towards animals. *Anthrozoos, 4,* 184–191.

Kalof, L. (2000). The multi-layered discourses of animal concern. In H. Addams & J. Proops (Eds.), *Social discourse and environmental policy: An application of Q methodology* (pp. 174–195). Cheltenham, U.K.: Edward Elgar.

Kheel, M. (1995). License to kill: An ecofeminist critique of hunters' discourse. In C. Adams & J. Donovan (Eds.), *Animals & women: Feminist theoretical explanations* (pp. 85–125). Durham, N.C.: Duke University Press.

Lawrence, E. A. (1995). Cultural perceptions of differences between people and animals: A key to understanding human–animal relationships. *Journal of American Culture, 18,* 75–82.

Lerner, J., & Kalof, L. (1999). The animal text: Message and meaning in television advertisements. *Sociological Quarterly, 40,* 565–586.

Luke, B. (1998). Violent love: Hunting, heterosexuality, and the erotics of men's predation. *Feminist Studies, 24,* 627–655.

Mason, J. (1998). *An unnatural order: Why we are destroying the planet and each other.* New York: Continuum.

McLaughlin, P. (1998). Rethinking the agrarian question: The limits of essentialism and the promise of evolutionism. *Human Ecology Review, 5,* 25–39.

Michel, S. (1998). Golden eagles and the environmental politics of care. In J. Wolch & J. Emel (Eds.), *Animal geographies: Place, politics and identity in the nature-culture borderlands* (pp. 162–183). New York: Verso.

Mullin, M. H. (1999). Mirrors and windows: Sociocultural studies of human–animal relationships. *Annual Review of Anthropology, 28,* 201–224.

Opotow, S. (1993). Animals and the scope of justice. *Journal of Social Issues, 49,* 71–85.

Ortega y Gasset, J. (1972). *Meditations on hunting.* New York: Scribner (Original work published 1942).

Osgood, C., Suci, G., & Tannenbaum, P. (1957). *The measurement of meaning.* Urbana: University of Illinois Press.

Paul, E. S. (2000). Empathy with animals and with humans: Are they linked? *Anthrozoos, 13,* 194–202.

Peek, C. W., Dunham, C. C., & Dietz, B. E. (1997). Gender, relational role orientation and affinity for animal rights. *Sex Roles, 37,* 905–921.

Plous, S. (1993). Psychological mechanisms in the human use of animals. *Journal of Social Issues, 49,* 11–52.

Plumwood, V. (1993). *Feminism and the mastery of nature.* London: Routledge.

Ritvo, H. (1995). Border trouble: Shifting the line between people and other animals. *Social Research, 62,* 481–499.

Shepard, P. (1996). *The others: How animals made us human.* Washington, D.C.: Island Press.

Shipman, P. (1994). *The evolution of racism: Human differences in the use and abuse of science.* New York: Simon and Schuster.

Snyder, G. (1969). *Earth house hold: Technical notes and queries to fellow Dharma revolutionaries.* New York: New Directions.

Spiegel, M. (1996). *The dreaded comparison: Human and animal slavery.* New York: Mirror Books.

9

Trees and Human Identity

Robert Sommer

When I was approached 15 years ago by the USDA Forest Service to survey residents' attitudes toward city trees, I wasn't certain that most people had clear opinions. It was possible that trees, like lampposts and fire hydrants, were street furniture passed by unnoticed. Our interviews with city dwellers quickly disabused us of this naive notion. We received responses like this:

"The Chinese pistache on our street makes it attractive to us and a prime choice of location when we bought the house."

"The community looks very good with trees in all the yards. When one is cut down, it is not the same, even if another tree is planted in its place."

"A city without trees is like a day without the sun."

Personal pronouns frequently preceded a species name in written comments; e.g., "our elm tree," as distinct from others' elm trees. Street trees were viewed as an integral component of the house and lot, the neighborhood, and sometimes of the city. The return rate on mailed questionnaires was surprisingly high, indicating a strong interest in the topic. Mailings to street addresses without family names produced returns in excess of 50 percent in repeated surveys (Sommer, Guenther, & Barker, 1990). Spontaneous comments thanked us for undertaking the survey, and people alerted us to gaps in city maintenance, or reported that a favorite tree was ill or injured. We relayed these comments to the appropriate city agency.

Trees are more than a decorative feature of the landscape. Many street and city names are based upon tree names. All this helps to shape individual and collective identities. The family tree is both a metaphor and a possession. Connections between trees and human identity can be indirect, as in trees enhancing home and neighborhood and raising property

Table 9.1
Identity effects of trees as seen in different lines of research

Physical factors	Tree canopy affects air quality, temperature, wind speed, noise, water runoff, and other natural processes that may influence human health and well-being.
Aesthetic factors	Trees make homes and neighborhoods more desirable, thereby enhancing individual and community self-images. Conversely, tree loss can produce grief responses, reflecting a diminution of self.
Economic factors	Trees add to the value of homes and neighborhoods, and this has a positive effect on self-image
Social factors	The presence of trees can improve neighborhood interaction. People identify more with trees they have planted themselves. Organized planting and maintenance programs lead to individual and collective empowerment.
Psychological factors	In both self-report and physiological studies, contact with greenery has restorative value. This can restore equilibrium to a person's relationship to the natural environment and heal a damaged self.

values, thereby contributing to the residents' self-images. As table 9.1 indicates, trees not only make economic and physical contributions to human well-being and sense of self, they also contribute in aesthetic, social, and psychological ways. There is something deeper, spiritual, and almost ineffable about people's attachment to trees (figure 9.1) (Schroeder, 1991).

Tree loss is another area where identity issues are apparent. Some of our respondents made us aware of the psychological outcomes of tree loss, which seemed similar in form to the grief accompanying the death of a family member. Samuels (1999) describes the responses of farmers in Colorado who lost most of their elm windbreaks to Dutch elm disease: "You want to cry, but you are too damn big." "It's like losing a kid." In 1989 hurricane Hugo came inland and destroyed much of the urban forest in Charleston, South Carolina. The following spring and summer, Hull (1992) undertook a telephone survey among Charleston residents.

Figure 9.1
Children feel a special affinity for trees that can be developed through environmental education.

More than 30 percent identified the trees as the most significant single feature of the city that was damaged. The respondents gave detailed explanations of how trees had given Charleston a special ambience, beautified and differentiated neighborhoods, and provided benefits such as shade, energy conservation, increased property values, and tourism. People also mentioned the spiritual dimensions of trees, how trees expressed the beauty of life, hope, God's work, and the belief that nature nourishes the soul. They saw the trees as an indication of civic concern, providing connection to the past, both in terms of personal memories and family history.

Efforts on Behalf of Urban Trees

The focus of this chapter is on trees in urban areas rather than forest trees. In urban areas, environmental organizations use tree planting as a means to build local identity, turning a street of strangers into a community. One activist admitted, "We don't know if we're organizing communities to plant trees or planting trees to organize communities" ("Forest Service," 1996, p. Y21). The strongest community building occurs in neighborhoods that focus their efforts on common spaces rather than yard planting (McLain, 1996). Trees create a canopy over residential streets, putting a "roof" over a neighborhood, forming natural bridges that unite two sides of a street.

City trees are heavily dependent upon people's actions for their survival. Those who maintain city trees have concluded that the social milieu surrounding the tree is as important as the physical milieu (Sklar & Ames, 1985). Discussing the national decline in maintenance budgets for municipal trees, Kay (1976) concludes, "Our city trees are in trouble because our cities are in trouble" (p. 21), adding that "You will find no tree-mourner like the apartment dweller watching the death of a tree before the front door. The urban-dweller's pain is personal" (p. 22). Adopt-a-tree programs enlist local residents to care for individual trees, attempting to develop a personal relationship by giving the tree a specific identity. When new trees were planted in a Boston neighborhood, pamphlets appeared in mailboxes announcing, "I've just moved in, actually, and there are a few things you should know about me. There are some things you can do to keep me alive and well. You should make sure I get 10 gallons of water a day. But don't drown me. Keep my base neat and weeded. . . . Don't salt near me in winter" (Kay, 1976, p. 26). Given the precarious situation of city trees, there seems nothing odd about this plea from a tree to its human neighbors (figure 9.2).

Harmful Aspects of Urban Trees

Although most of this chapter concerns their benefits, city trees can be major liabilities when strong winds or ice storms bring them crashing down on people, houses, roads, parked cars, and power lines. Even during tranquil times, not all tree species are equally appreciated by city residents. Some are regarded as messy and dirty (always dropping something) or as dangerous (falling limbs or detritus that people can trip over); some have roots that disrupt the sidewalk or sewer lines or branches that interfere with power lines; and some release allergenic pollen. Birds and squirrels roosting in trees are considered by some residents to be an amenity but by others to be a nuisance. In high crime areas, trees may block light and make a neighborhood seem less safe to pedestrians (Schroeder & Anderson, 1984). Some traffic engineers refer to city trees as FHOs (fixed hazardous objects) that impede the smooth flow of traffic (Duany & Plater-Zyberk, 1992). The relationship between humans and trees is not without ambivalence.

Having summarized some of the ways that trees are important in people's lives, the next sections of this chapter describe research findings and theories about the psychological significance of trees, with special attention to identity issues. A final section discusses the implications of the special bonds between people and trees for environmental theory, research, and practice.

Research on Tree-Planting Programs

Because city trees are so dependent on human assistance during their early years, urban foresters have concluded that the social environment around a tree is as important as the physical environment for ensuring early tree survival and well-being (McBride & Beatty, 1992). Sklar and Ames (1985) report higher survival for parkway trees planted in neighborhoods through block parties than for trees planted without

Figure 9.2
People organize to save city trees and they feel grief when one is lost.

community participation. Dwyer and Schroeder (1994) believe that participation in planting programs creates a stronger sense of community, empowers inner-city residents to improve neighborhood conditions, and promotes environmental responsibility and ethics (figure 9.3a–c) (see chapter 10). Residents learn they can choose and control the conditions of their environment. The national research agenda for urban forestry in the 1990s (ISA, 1991) considered community involvement to be critical for the continued vitality of the urban forest.

The past two decades have brought extensive multimethod, interdisciplinary documentation of the benefits of city trees, and there are some excellent reviews of this research (e.g., Dwyer & Schroeder, 1994; Kaplan & Kaplan, 1989; Ulrich, 1993). The findings include the following:

• Physical scientists have shown that the urban tree canopy moderates temperature and lessens the heat island effect; reduces wind speed, noise, and water runoff; and contributes to energy conservation. They have developed global models on forests' effects on carbon dioxide emissions, and economists have translated these findings into monetary benefits (Bolund & Hunhammar, 1999; McPherson, Simpson, Peper, & Xiao, 1999).

• Trees increase the property values of unimproved lots and homes (Martin, Maggio, & Appel, 1989; Payne & Strom, 1975), and homes are the largest single investment and capital asset for many families.

• Behavioral scientists have shown the aesthetic, stress-reducing, and restorative effects of trees and other greenery in a variety of environmental contexts for adults and for children (Kaplan & Kaplan, 1989; Kuo, Bacaicoa, & Sullivan, 1998, Ulrich et al., 1991).

• It has been shown repeatedly, both in simulation studies and in community surveys, that well-treed streets are preferred over those with few trees (Schroeder & Cannon, 1983).

My research team explored ways of increasing public involvement with city trees, first in planting, and then in maintenance. The planting studies used a replicated design in three California cities. In each location, we compared the attitudes of householders who planted front-yard trees themselves, either as part of a community program or on their own, with residents in the same neighborhood whose trees had been planted by the city or by a developer (Sommer, Learey, Summit, & Tirrell, 1994). The results showed that residents who planted trees themselves were

(a)

(b)

Figure 9.3
(a–c) Community tree plantings. Tree-planting programs empower and educate residents and enhance their sense of community.

(c)

Figure 9.3 (a–c) (continued)

more satisfied with the following: how the tree was staked or supported at planting time, the location selected for the tree, tree maintenance, the perception that the tree improved the yard or neighborhood, and the species planted. They also had a reduced desire to have the tree removed or replaced. There were additional benefits for those residents who had been part of a community shade tree program: increased mutual assistance during planting, becoming better acquainted with neighbors, greater access to maintenance information and technical assistance, and more willingness to use a telephone information service operated by a voluntary organization for future maintenance problems (Nannini, Sommer, & Meyers, 1998).

Because program membership and tree planting were correlated ($r = 0.34$, $p < .01$) while at the same time each correlated with overall satisfaction with the tree, partial correlations were used to identify the relative contributions of each to overall satisfaction. When program affiliation was correlated with overall satisfaction, holding planting participation constant, the resulting partial coefficient was nonsignificant, $r = 0.08$. When planting participation was correlated with overall satisfaction, with program affiliation held constant, the partial coefficient was

0.31 (p < .01). The clear implication of the partial coefficients is that planting the tree oneself is a more important contributor to resident satisfaction with the front-yard tree than being part of a community program.

Nannini, Sommer, and Meyers (1998) looked at the use of volunteers to inspect street trees for Dutch elm disease. Dutch elm disease is a significant threat to the elm population in many American cities. Since there is no effective treatment for the disease, the best available method is careful monitoring followed by quick removal of infected trees. Owing to shrinking municipal budgets, resident participation in community programs provides a way to economically maintain urban services. An alliance of professionals and volunteers is potentially more cost-effective and successful in monitoring tree condition than either group operating alone. A systematic comparison in Brookline, Massachusetts, found that tree data collected by trained volunteers were valid and their accuracy compared favorably with levels found among a control group of certified arborists (Bloniarz & Ryan, 1996).

In the early 1990s the first case of Dutch elm disease was detected in Sacramento County, California. The Sacramento Tree Foundation, supported by the city council and local businesses, started the Save the Elms Project (STEP) (figure 9.4). The municipal utility distributed 20,000 brochures in neighborhoods containing a lot of elms, asking for volunteers to conduct periodic tree inspections. There were 269 residents who initially expressed interest in joining the program, but owing to scheduling conflicts and time constraints, only half of them received training. The participants were given a full day of training, after which they adopted areas in their neighborhood containing an average of thirty-five elm trees each. Throughout the year, but most intensively during spring and summer, the volunteers examined their trees and reported on their condition to the Sacramento Tree Foundation. The participants got to know individual trees, which helped in detecting the changes that occurred at the onset of disease (Gemmel et al., 1995). Since the program began, the number of newly infected trees each year has steadily decreased, from forty in 1993 to eight infected trees in 1996.

We undertook a survey in 1996 to investigate the effects of community participation on volunteers' knowledge of their urban forest and attitudes toward their community. The survey was mailed to the same

Figure 9.4
Recruiting volunteers for Save the Elms Program.

269 residents who expressed initial interest in STEP. The first section of the questionnaire contained a series of statements that indirectly addressed individuals' sense of community and sense of empowerment. The second section contained questions from Gruber and Shelton's (1987) Resident Satisfaction Scale. The third section included nine multiple-choice questions that assessed individuals' knowledge of Dutch elm disease and the local urban forest.

More than 85 percent of those who volunteered for the community effort indicated that they were motivated by their concern for Sacramento's trees. Of the reasons provided by nonparticipants, only 9 percent cited a lack of interest in contributing to the efforts of the community project, while the majority named scheduling difficulties and time constraints as the primary obstacle to participation in STEP. A one-way analysis of variance indicated that individuals who participated in STEP were significantly more pleased with their neighborhoods than those who did not participate; $F(1, 117) = 4.75, p < .05$. Volunteers were significantly more knowledgeable about Sacramento's urban forest than individuals who did not participate in STEP; $F(1, 118) = 52.24, p < .001$.

Volunteers answered an average of 5.10 (SD = 2.05) questions correctly, compared with an average of 2.59 (SD = 1.68) answered correctly by those who did not participate in STEP.

There was more neighbor assistance among participants in tree-planting programs than among nonparticipants; this in turn led participants to become better acquainted with their neighbors (Sommer, Learey, Summit, & Tirrell, 1994). Observations in several cities revealed instances of neighbors meeting for the first time during tree planting (Learey, 1994). Interactions among neighbors in our study included such mundane activities as borrowing or lending tools, informal visits, and asking for assistance. Through these interactions, neighbors provided each other with instrumental, personal, and informational support.

Participation in the tree-planting programs also appeared to increase community ties, which form the basis for a sense of community identity. Again, the relationship is reciprocal: programs of neighborhood enhancement, including tree planting and maintenance, increase identification with the neighborhood (Rohe, 1985). As people identify with their neighborhoods, they personalize their homes, which increases community ties (Brown, 1987; Taylor, 1988). In turn, community ties will facilitate projects to improve the neighborhood environment.

Theories on the Psychological Significance of Trees

The human species developed over millennia in contact with trees; it was both influenced by trees and exerted a major influence on tree growth patterns. Trees are extolled in myth, song, poetry, and religion. These historical connections are expressed today in feelings of kinship, protectiveness, and mutual benefit. People identify with trees in general and with specific trees associated with events, people, and neighborhoods. A number of complementary theories have been developed concerning the relationship between people and trees. As indicated in table 9.2, Darwinian approaches (Quantz, 1897), depth psychology (Jung, 1979), phenomenological approaches (e.g., Davies, 1988), affordance theory (Gibson, 1979), and ecopsychology (Roszak, 1992) elaborate on the psychological significance of trees for people.

Table 9.2
Theories of the psychological significance of trees

Darwinian Approaches (S. R. Kellert & E. O. Wilson; J. O. Quantz)	The role of trees in natural selection has influenced latent and manifest preferences. This theory looks more at approach and avoidance rather than identity per se, although preferences are part of self-image (I am what I like; I like what I am). Myths of belief that people who were created from trees or were transformed into trees have influenced cultural practices, are embodied in stories and song, and find expression in children's animistic perception of trees. The life of a specific individual and a specific tree can be intertwined.
Depth psychology (C. G. Jung)	With a focus on parallels between human and arboreal development, the tree is seen as an archetype in the human collective unconscious. Personality tests such as the H-T-P use tree drawing to investigate identity issues.
Phenomenological approaches (Davies, 1988; Fulford, 1995; Tuan, 1979; Altman, 1993)	Relies heavily on metaphor between the natural and the human world. Roots, trunk, and canopy mirror the infernal, earthly, and heavenly domains, respectively; other features such as flowers, fruit, and color supply subsidiary themes relating trees to people and society.
Affordance theory (J. J. Gibson)	Real-world perception is shaped and refined through interaction with the outside world. During first-hand encounters, individuals learn properties of objects and their own place in the world. The theory gives more emphasis to perceptual learning than to identity.
Ecopsychology (T. Roszak)	Beyond the individual self, there is an ecological self that is nurtured through contact with and concern for the natural environment. A person should feel at one with nature, and if these feelings are absent or distorted, a healing process is needed.

Darwinian Approaches

J. O. Quantz (1897), an early developmental psychologist, was heavily influenced by both Darwinian functionalism and the anthropology of his day. Quantz suggests that a belief in tree spirits is as old as human civilization and traces religion to sacred groves, observing that "the groves were God's first temples" (p. 471). The sound of rustling leaves was interpreted as the oracular speech of tree spirits (p. 480), and the divining rod used for detecting water derives from the sacred tree with its magic powers. Even today, some wilderness managers describe themselves as "keepers of the sacred grove." Quantz considered the common practice of planting and dedicating trees to the memory of heroes and great events to be a modern residue of dendrolatry (tree worship).

Quantz describes the special characteristics of the Life Tree. Variations of this concept, central to concepts of human identity, including the World Tree and Tree of Paradise, appear in accounts of sacred trees (e.g., Altman, 1993 and Davies, 1988). The Life Tree is host to a spirit that establishes relationships with humans. In some instances, humans are created from these trees; in other instances, mortals are transformed into trees. From these two concepts—creation from trees and transformation into them—arises the more specific notion of a sympathetic connection between the life of a person and that of a particular tree. Hill (2000) provides a contemporary illustration of the bond formed between an individual and a specific tree. Some of today's tree-planting organizations sponsor Memory Tree programs, in which family members plant and maintain trees in honor of a departed loved one. In some cultures, trees are planted at the birth of a child and twin trees at a marriage. There is an obligation for the individual or couple to look after the birth tree or marriage trees and a perception that the fate of the individual and the tree, as a green alter ego, are intertwined. An ancient form of medical treatment was the supposed transfer of the ailment from person to tree. If the illness could be passed into a tree, the person would be relieved.

Quantz was given access to interviews with children and teachers conducted by the noted developmental psychologist G. Stanley Hall (1904). In these interviews, the children ascribed anthropomorphic qualities to trees, including reasoning, intelligence, emotions, and morality. Trees possessed feelings and suffered when they were trimmed or cut (see also chapter 5). The children found parallels between the tree and the human body: the limbs, trunk, and roots are its arms and feet, the leaves its clothing, the bark its skin, from which sap oozed as blood when the tree was injured. They reported that trees reciprocate their own feelings; they like to have little children around and will cry when lonesome, make shade just for little boys and girls, watch over the house, and talk to children who can understand what trees are saying. Some children considered all trees to be good, but the majority restricted goodness to those providing specific benefits, such as shade, fruit, or protection to birds. Some children expressed sympathy for crooked trees, but others considered deformity to be retribution for bad actions. They believed that trees put on new dresses during seasonal changes and are ashamed when all their leaves fall off. While the children saw individual trees as friends, based on affordances, forests tended to be associated with darkness and danger, a threat to identity, as in becoming lost or swallowed up by a forest or attacked by a wild animal.

More contemporary manifestations of Darwinian approaches to people–tree relationships are found in Kellert and Wilson's (1993) biophilia theory, Appleton's (1990) prospect-refuge theory, and Orians's (1986) savanna hypothesis. According to the concept of biophilia, defined as the innate emotional affiliation of humans with other living organisms, many human preferences were shaped over millennia through interactions with features of the environment that were helpful to the survival of the species in its early development (Kellert & Wilson, 1993). Through a process of gene-environment coevolution, the multiple strands of environmental stimuli become part of human culture. This allows a rapid adjustment to environmental changes through adaptations invented and transmitted without precise genetic prescription (Wilson, 1998). Even when humans are removed from the stimuli originally provoking an emotional response, biophilia theory maintains that the connections remain in latent form and find expression as preferences and aversions.

Landscape preference studies have shown that most people prefer natural, verdant, and open landscapes to those that are built, dry, or enclosed. The desired landscapes are assumed to be those which evolution has "taught" our species to be beneficial and worthy of approach rather than being avoided (Hull & Revell, 1989). The appearance of a tree evokes expectations about the fecundity of its host environment, helping to answer the question: "Is this a good place to stay?" Appleton (1990) describes desirable trees as offering prospect (for seeing prey or giving early warning of a predator's approach) and refuge, both of which provided the evolutionary advantages to early humans of seeing without being seen. Orians's (1986) savanna hypothesis sees preferred trees species as those associated with optimal habitats in the African savanna, where humankind spent its early millennia. These trees are more broad than tall and have canopies that are more wide than deep. A preference for species such as the acacia, which indicated desirable environments with ample water and game possibilities, became ingrained as a biophilic response through a lengthy process of gene-culture coevolution. Support for the savanna hypothesis is seen in landscape painting, cultivation practices, and preference studies among samples on many continents (Heerwagen & Orians, 1993; Sommer & Summit, 1996; Sommer, 1997).

Depth Psychology

The tree as metaphor for the course of human life was a favorite theme of the psychoanalyst C. G. Jung (1979). The potential for the mature tree is present in the tiny seed just as the potential for an adult human is present in the fertilized ovum. Considering the tree to be an archetype, or innate pattern that is part of a collective unconscious common to all humans, Jung collected paintings of trees done by his patients. He considered the tree to be symbolic of the life course, with the roots as genealogy, the trunk as the evolving personal identity, the branches as those characteristics and traits that connect to the environment, and the fruit as the creative products of an individual life (Metzner, 1981). Jungian psychology blurs the inner-outer distinction in making a correspondence between the outer wilderness of nature and the "wilderness" of the unconscious mind (Schroeder, 1991). In a similar vein, shamanistic traditions describe the central axis of a human being as an inner tree. Traveling to this central axis, one encounters the axis mundi, or pillar of the world, the cosmic hub (Metzner, 1981).

Drawing a tree is part of the House-Tree-Person projective test used to assess deeper levels of personality (Buck, 1948). In the original version of this procedure, the person was requested to draw in succession a house, a tree, and a person. Hammer (1986) modified this technique by asking the participant to imagine being that house, that tree, or that person, and to speak for it. Metzner (1981) goes even further by asking his clients to "draw the tree of your life," which connects directly to identity. The drawn tree is seen as expressing the yearning to grow and move from earth to the heavens, making it an important metaphor for self-enfoldment and the building of personal identity. The branching symbolizes protection, shade, nourishment, growth, regeneration, and determination (Burns, 1987).

Jungian depth theories, because they are individual centered, speak directly to identity issues. Projective tests that ask the person to role play being a tree represent perhaps the ultimate in identity merger. The spiritual dimensions of trees fit well into the framework of depth psychology.

Phenomenological Approaches

Like psychoanalytic theories, phenomenological discussions of trees rely heavily on metaphor. Roots, trunk, and canopy mirror the infernal, earthly, and heavenly domains, respectively; other features such as flowers, fruit, and color supply subsidiary themes. Davies (1988) observed that "Trees are not simply good to climb, they are good to think. Much of their wood is fuel for metaphorical fires" (p. 34). He contrasted the importance of trees in metaphor with the infrequent references to grass, the most universal and successful of plants, which has not fed the flames of creative thought to an equal extent. Looking at the symbolism of trees in Georgian England, Daniels (1988) sees them as a stabilizing influence during a period of accelerating social change. He discusses the connotations surrounding particular species, such as the venerable and rich associations adhering to the oak and the patrician connotations of parkland trees as distinct from the more plebian and fast-growing firs, which were not nearly as celebrated in verse and story. While the oak sheltered all around it, the cedar was seen as destroying everything in its shade (Fulford, 1995). Others attribute the prominence of oaks in song and verse, not only to age and size, but also to the deep, thick roots, which convey a sense of permanence and mystery. This made them a suitable image for stable government. Edmund Burke, who was himself described as an oak in a poem by Wordsworth, depicted England's form of government in similar terms. Burke was opposed by radicals such as Thomas Paine, who saw the liberty tree of the French

Revolution as the new growth possible after an old unjust system is uprooted (Fulford, 1995). Settling in a place means putting down roots, perhaps by planting long-lived, deeply rooted trees like the oak.

Human attitudes toward forests, as distinct from individual trees, were originally mostly negative. Forests were seen as places of darkness and danger, the gathering places of dangerous animals, witches, and outlaws. In children's stories, the forest was not a place for a stroll or play; it represented threat and abandonment. However, unlike blizzards and floods, which might be seen as pursuing their victims, forests were dangerous only to those who encroached on their domain, especially those who violated their rules by injuring sacred trees or animals (Tuan, 1979). Residues of these fearful attitudes remain, as in the expression, "We are not out of the woods yet."

Altman (1993) attributes the deep kinship between people and individual trees to "a primary quality in common that is found in few other beings. We both share a vertical perspective"(p. 7). Some native peoples referred to humans as "walking trees," whose spine is the trunk, pelvis enfolds the roots, and brain is in the branches. Altman maintains that this similarity of posture was one of the factors leading people to see trees as natural friends and allies. The image of a tree has the capacity to raise human consciousness in expressing values such as permanence, stability, trustworthiness, fertility, and generosity. Theologian Martin Buber wrote, "I can look upon (a tree) as a picture . . . as movement . . . and I can classify it in a species and study it. . . . It can, however, also come about, if I have both will and grace, that in considering the tree, I become bound up in a relation with it" (quoted in "Holy hickory," 1994, p. 34).

Affordance Theory

J. J. Gibson (1979) developed a theory of real-world perception that emphasized its practical value for the individual, in terms of people seeing things as opportunities. Specifically in regard to trees, the theory focused on their practical benefits and uses, which are discovered through perceptual learning. Knowledge of their structural properties is innate in Gibson's theory, in terms of features such as roundness, hardness, and height that have survival relevance for the species. Individuals explore and refine this innate knowledge through perceptual learning, and these features become ingrained within a culture and transmitted. Through this combination of perceptual learning and social transmission, the individual learns, for example, how different trees should be climbed since some will have smooth bark while others will have rough bark, and some will have low branches for easy footing while others will have none.

For rural and suburban children, trees afford significant play opportunities (climbing, hide-and-seek, rope swings, tree forts). Children in inner-city areas have less direct access to trees, yet engage in more play activities and play more creatively when there are trees and grass available (Faber, Wiley, Kuo, & Sullivan, 1996). For adults, trees offer the means to satisfy basic needs connected with shelter, fuel, food, and medicine. In modern urban life, trees confer many benefits, including shade, aesthetic enhancement, energy conservation, reduced air pollution, privacy, reduced noise, seasonal markers, wildlife habitat, and enhancement of property values. Identity figures indirectly in these benefits. Within the larger gestalt of person-home-neighborhood, trees are not so much a part of an individual's identity as a part of the environment on which survival of the individual depends. The loss of trees is therefore a serious matter, in terms of the many benefits they offer, although the intensity of the response to tree loss in some cases may require explanations from depth psychology or ecopsychology theory. In his 1907 Arbor Day proclamation, President Theodore Roosevelt declared, "A people without children would face a hopeless future; a country without trees is almost as hopeless" (cited in Samuels, 1999, p. 101).

Ecopsychology Theory

Ecopsychology has roots in several earlier theories, including Gibson's ecological psychology with its emphasis on real-world perception; transpersonal psychology, which focuses on transcendental experience; and deep ecology, which rejects the idea of humans being at the top of the evolutionary pyramid (Fox, 1990). Ecopsychology assumes an ecological unconscious that is similar in many respects to Jung's collective unconscious, but with greater emphasis on the physical environment, not merely as symbol, but as a focus of concern (Roszak, 1992). Following the Gaia hypothesis (Lovelock, 1990), ecopsychology maintains that the planet Earth is a single living organism and there should be no distinction between living and nonliving matter. The latter view differs from biophilia, which stresses the special affinity between humans and other life forms, with physical features of the environment seen mainly as life supports.

For city residents, trees provide contact with natural rhythms and life forms, offering leaf color as seasonal markers; the gentle motion and sound of rustling leaves; filtered light; and habitat for birds, squirrels, and insects. Shade for outdoor activities is especially critical in cities subject to heat island effects because trees can moderate temperatures and save energy used for air conditioning. Ecopsychology speaks directly to identity issues, mainly by broadening the concept. Calling the field "transpersonal ecology," Fox (1990) defines it as the study of the ecological self *beyond the human identity* (italics mine). He maintains that humans are overly homocentric and need to lose their sense of self-importance among species. In place of the conventional image of a ladder with humans occupying the top rung, he employs a tree metaphor to describe the separate branches of evolutionary development, with humans as but one leaf on the tree of life. Even though leaves fall and branches break, the tree itself will continue (Winter, 1996). Shedding homocentrism and adopting a broader view of self has emotional, perceptual, and spiritual implications. The concept of identification rather than merger is central to this shift. One feels identified with the planet, not that one is the planet. One becomes identified with trees, one does not become a tree. Winter (1996) eloquently describes this shift in awareness: "It is less about information and more about identification. Less of a decision and more of a dropping into a fuller experience of oneself. . . . Less about knowing and more about appreciating. The ecological self is an expanded, more gracious, more spacious sense of self" (p. 264).

Ecopsychology also has an action component, in that healing the planet is seen as a way of healing the self. Winter (1996) defines the goal of the field as creating a sustainable world. Involvement in environmental causes is a spiritual exercise that is beneficial to the self. Planting trees is good for the psyche and for the environment. As the ecological self expands within the person, environmentally destructive consumerist values will lose importance.

All of these theories can be used to understand the implications of trees for human identity. As depicted in table 9.3, these theories present three

Table 9.3
Basis of human identification with trees

---

1. Physical and metaphorical resemblance
   Both are vertical, alive, have growth cycles, similarity of parts (canopy =
   head, trunk = body, branches = arms, roots = feet)
2. Myths and legends
   People created from trees, people turned into trees, tree spirits, sacred
   groves, haunted forests
3. Intertwined fate
   People dependent upon trees for shelter, fuel, food, building materials,
   prospect, medicines, shade, aesthetic pleasure, contact with nature, wildlife
   habitat, children's play, restorative qualities, seasonal markers, energy
   conservation, windbreaks, enhanced property[a]
   Trees dependent on people for planting, care, protection, monitoring for
   disease, removal of diseased trees, research, silviculture[b]

---

[a] Includes benefits for humans during both evolutionary history and in modern
times.
[b] Refers to trees near human settlements that need to be planted, maintained,
and protected by humans.

main bases for people's identification with trees: physical and metaphorical resemblance, myth and legends, and their intertwined fate.

Implications for Environmental Policies, Practices, and Research

The two major themes of this book, nature and identity, are large, fuzzy
concepts on which it is difficult to conduct empirical studies. Definitions
of both concepts are vague, imprecise, nonconsensual, and often ad hoc.
It is much easier to investigate the amount of time people spend gardening or hiking than to study their beliefs and actions in relation to
nature in the abstract. I did not start out studying nature or identity. I
conducted research on community tree-planting programs and later, preference for generic tree shapes, which brought me to theories of identity
and nature. The research came first, with the theories used to interpret
and extend the findings. I doubt I would have received funding from the
USDA Forest Service if I had proposed to study "trees and identity." This
federal agency must preserve its credibility in a highly contested political arena, where many groups regard its activities with suspicion. I

studied narrow, operationally defined topics and published empirical papers in refereed technical journals.

This chapter, the first that connects my research with identity issues, suggests that the psychological significance of trees has implications for urban environmental programs. Trees should be part of neighborhood improvement projects, with the provision that residents should be involved in tree selection, planning, and maintenance. Professionals can contribute technical expertise without taking "ownership" of local trees. Expanding the opportunities for resident involvement in tree monitoring and care should be explored. Although there are cross-national preferences for certain generic tree forms, particularly the oak and acacia shapes, people are also likely to identify with familiar local trees (Sommer, 1997).

More research on this topic would help to inform theory and practice. There are several research approaches that can help us better understand how people relate to trees:

• Simulation methods can be used to gauge response to tree sickness, injury, death, and removal. Existing research on tree loss has been largely anecdotal, after the fact, and confounded with larger events, such as infestations and natural disasters.

• To supplement qualitative post hoc studies, there is a need for more systematic procedures, such as slide simulation with semantic differential ratings, and role-playing exercises.

• Most simulation studies have investigated the potential effects of adding trees to the landscape; much less has been directed toward the effects of tree loss. This line of research would help city officials respond appropriately to protests about tree removal, and point the way to effective ameliorative and replacement strategies. Should there be an immediate replacement planting or should the space be left empty for a time to allow mourning? How can residents be directly involved in replacement planting for a lost tree? How should a city arborist behave when it becomes necessary to remove trees that are part of home and neighborhood identity?

• There has been no longitudinal research on individuals and their birth trees. The practice of planting trees at a child's birth is losing ground with urbanization, but examples still occur that should be followed up. This can include imagined responses to vandalism of a birth tree, as in a stranger carving initials on it (presumed appropriation).

• Nor has there been research on family identification with memory trees, i.e., those planted, either on family or public space, to honor a departed family member. Research can be done using lists kept by organizations sponsoring memory tree programs.

• There are interesting identity issues in regard to trees linked to significant individuals. American Forests, a nonprofit conservation group, sells trees that are the direct descendents of living trees planted by famous individuals and events. Among their most popular offerings are seeds or cuttings from a tulip poplar growing on the woodlots of Mount Vernon, a weeping willow from the Graceland estate of Elvis Presley, and a sycamore from the Antietam battlefield in Maryland (Associated Press, 2001). What is the symbolic meaning of planting and caring for a tree associated with the life of a named individual?

In conclusion, theories of human contact with nature should recognize the special characteristics of trees and their role in fostering individual and community identity. As Brush and Moore (1976) pointed out, a major challenge for behavioral scientists involved in greening programs is to identify those aspects of city trees liked and disliked by residents. For city dwellers, trees are important parts of "near nature," and opportunities for tree monitoring and care by children as well as adults should be expanded.

References

Altman, N. (1993). *Sacred trees*. San Francisco: Sierra Club Books.

Appleton, J. (1990). *The symbolism of habitat*. Seattle: University of Washington Press.

Associated Press. (2001). Planting a tree can be a historic moment. Berkeley, Calif., *Daily Planet*, Mar. 23, p. 14.

Bloniarz, D. V., & Ryan, H. D. P. (1996). The use of volunteer initiatives in conducting urban forest inventories. *Journal of Arboriculture, 22*, 75–82.

Bolund, P., & Hunhammar, S. (1999). Ecosystem services in urban areas. *Ecological Economics, 29*, 293–301.

Brown, B. B. (1987). Territoriality. In D. Stokols & I. Altman (Eds.), *Handbook of environmental psychology*, vol. 1 (pp. 505–531). New York: Wiley.

Brush, R. O., & Moore, T. A. (1976). Some psychological and social aspects of trees in the city. In F. S. Santamour (Ed.), *Better trees for metropolitan landscapes*. USDA Forest Service Technical Report NE 22, pp. 25–29, Broomall, PA: USDA.

Buck, J. N. (1948). The H-T-P test. *Journal of Clinical Psychology, 4*, 151–159.

Burns, R. C. (1987). *Kinetic-house-tree-person drawings*. New York: Brunner/Mazel.

Daniels, S. (1988). The political iconography of woodland in later Georgian England. In D. Cosgrove & S. Daniels (Eds.), *The iconography of landscape* (pp. 43–80). Cambridge: Cambridge University Press.

Davies, D. (1988). The evocative symbolism of trees. In D. Cosgrove, & S. Daniels (Eds.), *The iconography of landscape.* Cambridge: Cambridge University Press.

Duany, A., & Plater-Zyberk, E. (1992). Second coming of the American small town. *Wilson Quarterly, 26,* 19–50.

Dwyer, F. F., & Schroeder, H. W. (1994). The human dimensions of urban forestry. *Journal of Arboriculture, 92,* 12–16.

Faber, A. J., Wiley, A. R., Kuo, F. E., & Sullivan, W. C. (1996). Children in the inner city: Nature as a resource for play. Paper presented at the Twenty-Seventh Annual Meeting of the Environmental Design Research Association.

Forest Service tries to gain an urban foothold. (1996). *New York Times,* Sept. 22, p. Y21.

Fox, W. (1990). *Toward a transpersonal ecology.* Boston: Shambhala.

Fulford, T. (1995). Cowper, Wordsworth, Clare: The politics of trees. *John Clare Society Journal, 14,* 1–7.

Gemmell, B., Fenkner, A., Ferri, M., & Williams, S. (1995). Stewardship of the mature elm tree canopy in Sacramento through citizen participation. Paper presented at the Conference on Benefits of the Urban Forest, Sacramento Tree Foundation.

Gibson, J. J. (1979). *The ecological approach to visual perception.* Boston: Houghton Mifflin.

Gruber, K. J., & Shelton, G. G. (1987). Assessment of neighborhood satisfaction by residents of three housing types. *Social Indicators Research, 19,* 303–315.

Hall, G. S. (1904). *Adolescence.* New York: Appleton.

Hammer, E. F. (1986). Graphic techniques with children and adolescents. In A. I. Rabin (Ed.), *Projective techniques for adolescents and children* (pp. 239–263). New York: Springer-Verlag.

Heerwagen, J. H., & Orians, G. H. (1993). Humans, habitats, and aesthetics. In S. R. Kellert & E. O. Wilson (Eds.), *The biophilia hypothesis* (pp. 138–172). Washington, D.C.: Island Press.

Hill, J. B. (2000). *The Legacy of Luna.* San Francisco: Harper.

Holy hickory and sacred sycamore. (1994). *Utne Reader, 61* (January) p. 34.

Hull, R. B. (1992). How the public values urban forests. *Journal of Arboriculture, 18,* 98–101.

Hull, R. B., & Revell, G. R. B. (1989). Cross-cultural comparison of landscape scenic beauty evaluations: A case study in Bali. *Journal of Environmental Psychology, 9,* 177–191.

ISA. (1991). A national research agenda for urban forestry in the 1990s. Urbana, Ill.: International Society of Arboriculture.

Jung, C. G. (1979). *Word and image*. Princeton, N.J.: Princeton University Press.

Kaplan, R., & Kaplan, S. (1989). *The experience of nature*. New York: Cambridge University Press.

Kay, J. H. (1976). The city tree. *Horticulture, 54,* 21–28.

Kellert, S. R., & Wilson, E. O. (Eds.). (1993). *The biophilia hypothesis*. Washington, D.C.: Island Press.

Kuo, F. E., Bacaicoa, M., & Sullivan, W. C. (1998). Transforming inner city landscapes. *Environment and Behavior, 30,* 28–59.

Learey, F. (1994). Themes of community participation in tree planting. In M. Francis, F. Lindsey, & J. S. Rice (Eds.), *The healing dimensions of plant-people relations* (pp. 383–389). Davis, Calif.: University of California, Davis Center for Design Research.

Lovelock, J. (1990). *The ages of Gaia*. New York: Bantam.

Martin, C., Maggio, R., & Appel, D. (1989). The contributory value of trees to residential property in the Austin, Texas metropolitan area. *Journal of Arboriculture, 15,* 72–76.

McBride, J., & Beatty, R. (1992). Comment. *Connections: Urban forestry research update, 1*(2), p. 1.

McLain, J. (1996). Reclaiming our communities in New Haven. *Urban Issues* (Yale School of Forestry), 7, p. 8.

McPherson, E. G., Simpson, J. R., Peper, P. J., & Xiao, Q. (1999). Benefit-cost analysis of Modesto's municipal urban forest. *Journal of Arboriculture, 25,* 235–248.

Metzner, R. 1981. The tree as a symbol of self unfoldment. *American Theosophist*. Fall issue, 289–300.

Nannini, D. K., Sommer, R., & Meyers, L. S. (1998). Resident involvement in inspecting trees for Dutch elm disease. *Journal of Arboriculture, 24,* 42–46.

Orians, G. H. (1986). An ecological and evolutionary approach to landscape aesthetics, In E. C. Penning-Rowsell & D. Lowenthal (Eds.), *Landscape meanings and values* (pp. 3–25). London: Allen and Unwin.

Payne, B. R., & Strom, S. (1975). The contribution of trees to the appraised value of unimproved residential lots. *Valuation, 22,* 36–45.

Quantz, J. O. (1897). Dendro-psychoses. *American Journal of Psychology, 9,* 449–506.

Rohe, W. M. (1985). Urban planning and mental health. In A. Wandersman & R. Hess (Eds.), *Beyond the individual: Environmental approaches and prevention*. New York: Haworth.

Roszak, T. (1992). *The voice of the earth*. New York: Simon and Schuster.

Samuels, G. B. (1999). *Enduring roots*. New Brunswick, N.J.: Rutgers University Press.

Schroeder, H. W. (1991). The spiritual aspect of nature. In *Proceedings of the Northeastern Recreation Research Symposium* (pp. 25–30). Radnor, Pa.:

USDA Forest Service Northeastern Forest Experiment Station, Report No. GTR-NE-160.

Schroeder, H. W., & Anderson, L. M. (1984). Perception of personal safety in urban recreational areas. *Journal of Leisure Research, 16,* 177–194.

Schroeder, H. W., & Cannon, W. N. (1983). The aesthetic contribution of trees to residential streets in Ohio towns. *Journal of Arboriculture, 9,* 237–243.

Sklar, F., & Ames, R. G. (1985). Staying alive: Street tree survival in the inner city. *Journal of Urban Affairs, 7,* 55–65.

Sommer, R. (1997). Further cross-national studies of tree form preference. *Ecological Psychology, 9,* 153–160.

Sommer, R., Guenther, H., & Barker, P. A. (1990). Surveying householder response to street trees. *Landscape Journal, 9,* 79–85.

Sommer, R., Learey, F., Summit, J., & Tirrell, M. (1994). Social and educational benefits of resident involvement in tree planting. *Journal of Arboriculture, 20,* 170–175.

Sommer, R., & Summit, J. (1996). Cross-national rankings of tree shape. *Ecological Psychology, 8,* 327–341.

Taylor, R. B. (1988). *Human territorial functioning.* Cambridge: Cambridge University Press.

Tuan, Y-F. (1979). *Landscapes of fear.* New York: Pantheon.

Ulrich, R. S. (1993). Biophilia, biophobia, and natural landscapes. In S. R. Kellert & E. O. Wilson (Eds.), *The biophilia hypothesis* (pp. 73–137). Washington, D.C.: Island Press.

Ulrich, R. S., Simons, R. F., Losito, B. D., Fiorito, E., Miles, M. A., & Zelson, M. E. (1991). Stress recovery during exposure to natural and urban environments. *Journal of Environmental Psychology, 11,* 201–230.

Wilson, E. O. (1998). *Consilience.* New York: Vintage.

Winter, D. N. (1996). *Ecological psychology.* New York: HarperCollins.

10

Identity, Involvement, and Expertise in the Inner City: Some Benefits of Tree-Planting Projects

Maureen E. Austin and Rachel Kaplan

Vacant lots rarely make a neighborhood look attractive. Passersby might make inferences about the residents—that they do not care, perhaps are not even aware of the physical appearance of their neighborhood. The residents may indeed have more pressing concerns than how their neighborhood looks to others. Alternatively, the vacant lots may be painful and constant reminders of the personal and communal deprivations of these residents. The context for this chapter is vacant lots in Detroit, Michigan, or more accurately, tree-planting projects that have transformed both these lots and the appearance of the neighborhoods. Along with these physical transformations came many other changes: citizens who engaged in community activities, people who learned from playing leadership roles, and individuals who came to have a new sense of who they are and what they can contribute. Thus the tree-planting projects serve to explicate the interplay among involvement, expertise, and identity.

It is not difficult to imagine that the transformation of overgrown vacant lots into well-kept community green spaces might affect how a neighborhood looks. Perhaps less self-evident is that it also changes how the residents come to see themselves. Our discussion draws on interviews with leaders in eleven Detroit neighborhoods as well as the lot keepers whose efforts sustained the projects for the months and years after the planting (Austin, 1999). From descriptions of their participation in organizing and maintaining neighborhood vacant lot projects emerged evidence of just how strong the connections between people and the physical setting can be. Indeed both place and people were transformed by the events.

Before turning to the interviews, we provide a brief historical overview of Detroit, its neighborhoods, its vacant lots, and the tree-planting program. This overview is useful in understanding present-day neighborhood settings. We then turn to insights gained from the interviews with leaders and lot keepers. Their reflections make tangible much that is invisible, both in terms of their own efforts and ideals, and in how the changes in the vacant lot also change the neighborhood and its residents. The interviews also tell us that not all efforts are successful; success, which is often the result of experience, is a key ingredient in forging pride and uplifting the residents.

Detroit: Then and Now

Detroit, "The Motor City," has long been identified with the automobile industry. The city has, however, seen major changes since the automobile assembly line first attracted many newcomers prior to World War I. From early on Detroit has had a reputation for being a city of neighborhoods. The steady flow of people moving to the city to work in its factories led to the creation of neighborhoods densely stocked with homes for working-class families.

This neighborhood structure persisted even with major changes in demographics after World War II. By then, as Darden, Hill, Thomas, and Thomas (1987) describe it, the decentralization of businesses and manufacturing as well as the increasing migration of middle-class residents to the suburbs were accompanied by a substantial immigration of Blacks from the South. In the process, the old neighborhoods changed radically, becoming areas where the "poorest of the poor" were struggling to make a living. Between the mid-1960s and 1990, Detroit lost well over one-quarter of its population as citizens left their homes and moved out of the inner city, leaving more than 65,000 vacant lots in the city (Grove, Vachta, McDonough & Burch, 1993).

Formerly known as "The City of Trees," Detroit was once home to hundreds of thousands of American elm trees, which added a graceful elegance to the city's streetscape. Not so long ago the city was full of neighborhoods teeming with families, living in homes packed side by side along streets lined with graceful trees. Today we find a starkly different scene. Falling victim to disease, old age, and development, many of the elm trees are long gone. Many neighborhood streets now have more vacant lots than homes, and more empty tree wells than trees. With city services strained by a shrinking budget, the lots receive little and only sporadic maintenance and quickly become overgrown, weedy sites of illegal dumping.

Vacant lots are not only an eyesore, they are also unsafe. Without regular maintenance, they turn into a sea of tall weeds that hide from view whatever may be taking place there (figure 10.1). A neighborhood resident, a participant in our study, shared her concern for the safety of children having to walk through these lots on the way to school:

There were weeds, or I should say grass that had grown about five feet tall. And I was at a meeting one time with the Chief of Police, and I told him you're not going to stop the crime and you're going to find bodies of children. We have five schools in our boundary and you're going to find children in [those lots]. But it was on the east side. It happened the very next year after that meeting. There was someone, a child, found in those weeds.

Despite the physical and social changes, a strong sense of neighborhood identity persists. Most neighborhoods have an established group or organization and local leaders who help ensure that neighborhood interests are heard and manifested in social advocacy programs such as

Figure 10.1
While some might consider this vacant lot in a southwest Detroit neighborhood a lovely setting filled with wildflowers, to a local resident it is a reminder of neighborhood decline, and a setting for illegal dumping and crime.

housing and development. Names for these neighborhood organizations sometimes correspond to their local street; in many instances, however, the name depicts the solidarity of its residents or a local cause, such as fighting blight or drugs. In many neighborhoods where local resident organizations are active, colorful signs bearing the neighborhood name are located at the neighborhood's entrance. Despite much hardship— poverty, crime, home demolition, and garbage-filled vacant lots—the residents identify with their neighborhood.

Neighborhood organizations can take it upon themselves to do something with the overgrown vacant lots in their neighborhood. Working with the city's forestry department and a local nonprofit tree-planting organization, neighborhood residents turn vacant lots into islands of green. A representative from the neighborhood group assumes a leadership role in this process, completing application paperwork and meeting with personnel from both the tree-planting organization and the city forestry department to ensure that the planting project details are attended to and local residents are kept informed. The leaders also sign a maintenance agreement, assuming responsibility on behalf of their organization for maintaining the vacant lot project after planting day.

The focus of this chapter is on tree-planting projects for vacant lots that took place between 1994 and 1997. Semistructured interviews were conducted with fourteen leaders and thirty-eight lot keepers. The leaders were asked to reflect on the planting process, their role in organizing local residents, use of the lot before and after planting day, lot maintenance efforts, and the neighborhood's sense of community. The lot keepers were asked about their participation in lot maintenance, as well as their perspective on lot use by neighborhood residents. While identity was not an explicit objective of the research, it emerged from the interview process as participants spoke of their involvement, experience, and the neighborhood. The quotes incorporated in our discussion, which were recorded during the taped sessions, provide a glimpse of the individual identities of almost all the leaders and many of the lot keepers.

Study Sites

The eleven neighborhood organizations included in this study are scattered throughout Detroit and serve different purposes. Four are concerned with neighborhood improvement and development, encompass-ing several blocks. Three are neighborhood block clubs involving a single block and approximately fifteen to thirty households. The remaining four are nonprofit community service agencies serving the needs of local res-idents. They include a men's homeless shelter, a community mental health agency, and two housing organizations.

There is no set form or structure to tree-planting projects for vacant lots. The neighborhood group or agency often identifies a special purpose or function for the lot, which is reflected in the finished project. Descrip-tions of three such projects illustrate how a completed project can be adapted to the neighborhood's needs.

An Eastside Block Club Pocket Park

This lot project is the result of countless hours invested over a number of years by members of a neighborhood block club on Detroit's east side. The site was an abandoned house that neighbors feared was unsafe and could provide a place for illegal activities. The residents of the neigh-borhood worked together to convince the city to demolish the house. In its place there is now a small pocket park nestled between two houses and cared for by members of the block club. A small white fence crosses the front of the lot and a winding brick pathway invites one to enter. In the center of the lot are benches facing one another so residents can sit and chat. The lot contains two shade trees, one flowering tree, numer-ous shrubs, and a sign bearing the block club's name. Since planting day the neighbors have wasted no time in making this site an integral part of their local neighborhood scene; numerous events have been held there, including barbecues, a retirement party, and a graduation party.

A Symbol of Neighborhood Improvement

On the city's north side is a planting project that is equally important to local residents yet serves a more formal function. Located along a busy thoroughfare transecting the neighborhood, the project is a symbol of neighborhood commitment to improvement and beautification. This large project consists of several contiguous lots bounded by two side streets. Topsoil was added to the site, giving it a slightly rolling topog-raphy. Railroad ties have been placed on either side of a wide woodchip pathway running diagonally through the lots, offering residents a place to sit when they visit the site. This lot also has a sign bearing the name of the neighborhood association. Instead of impromptu barbecues and social gatherings, formal events such as a dedication ceremony and association fundraisers are held here.

Beautifying a Neighborhood while Deterring Crime

A southwest Detroit neighborhood experiences considerable gang activity within its borders. The neighborhood has numerous vacant lots that are routinely used by local gang members as shortcuts for vehicles. The destructive use of these lots prompted a neighborhood nonprofit housing agency to sponsor a yearly lot beautification program. The vacant lot project consists of two adjacent lots, with a woodchip pathway leading through the lots and exiting into a small city park containing swing sets, a jungle gym, and a picnic table. Flowerbeds and trees have been planted strategically to each side of a metal pole gate in order to discourage entrance by vehicles. Discarded telephone poles placed horizontally along the rear border of the lots also prevent vehicles from coming in and serve as sitting areas. A community bulletin board placed at the front of the lots near the sidewalk is used to post notices of neighborhood meetings, garage sales, and other social events. The lots offer a beautiful entrance to the city park where neighbors can walk their dogs and watch their children playing safely in the park beyond (figure 10.2).

Neighborhood Leaders: Inspiration, Persistence, and Action

Planting day must be preceded by a great deal of planning and effort. The initiative taken by the neighborhood leader is critical to many phases of the process. While it is simple enough to make a telephone call and request an application for funds, planning and organizing the planting project requires a depth of commitment that many residents are not willing to make. The way in which leaders see themselves and relate to the neighborhood can have important ramifications for the outcome of the project.

With three projects having co-leaders, a total of fourteen individuals served in a leadership capacity for the eleven vacant lot planting projects. The majority (nine) of the leaders were African American. Four of the six agency leaders were male; seven of the eight neighborhood leaders were

Figure 10.2
With care and commitment, a vacant lot changes from a liability to a neighborhood park and focal point.

female. While they were not compensated directly for their leadership in the planting process, the agency leaders by virtue of their employment were paid for their efforts. The involvement of neighborhood leaders, on the other hand, was voluntary and they received no compensation.

Some of the vacant lot projects were the brainchild of the neighborhood leader. These were generally leaders who had successfully directed previous neighborhood improvement initiatives. In other cases the leaders were assigned this task, not unwillingly, as part of their service to the community. Regardless of prior experience with this type of neighborhood improvement project, the leaders seemed quite comfortable with their title and role because they had served in a similar capacity as either president of the local block club or neighborhood association, or director of a local nonprofit service agency.

Leaders Envision Community Identity
Leaders understand that it is difficult for many residents to believe their neighborhood can become transformed. It takes considerable foresight

to see what overgrown, weedy, garbage-filled lots might be transformed into, and then be able to apply skills, talents, and dedication to make that transformation happen. Two neighborhood leaders described the powerful impacts the local environment can have on how nearby residents perceive their neighborhood:

When you wake up and you turn the corner and you see trash, you see lots not cut, homes burned out and everything, you can't feel good about that. And so when you feel good about where you are, you do more.

The problem is that when you are used to, when you are growing up looking at garbage, you don't have any, you don't have no sense of what it could be. And I think people say yeah, it's such a dump, but they never did anything about it.

Many of the leaders shared descriptions of what they had envisioned at the onset of planning their neighborhood vacant lot project. Some leaders envisioned a neighborhood park, a green space, a place for families to picnic or where neighbors could visit with one another and interact with nature. Others envisioned the lots as gardens and sources of beauty for local residents. Both agency and neighborhood leaders shared their visions for the vacant lot project:

We could do our block sale over there, I mean I had wanted to see people use it, like maybe families come, they could picnic over there. There's a lot of things I had envisioned using that lot for.

Our goal was to put swings up there and to get a butterfly garden.

We wanted the park edible. With . . . an edible landscape, architecturally designed with fruit trees, boysenberry trees, raspberry trees, and everything that was in the park would be edible. I feel that the garden is a wholesome thing.

To help shape community identity, leaders need to see beyond the current circumstances and envision what is possible. The vision, however, is not enough; it must be articulated to local residents, and leaders must obtain the support necessary from residents to turn that vision into reality.

Neighborhood Leaders and Expertise

Organizing neighborhood planting projects requires diverse leadership qualities such as commitment, organization, and the ability to attend to details and navigate the city's bureaucratic channels. The leaders varied considerably in their styles, approaches, and strengths with respect to these qualities. Some leaders handled the details of project organization with ease; for others these were more perplexing. Some were close to the neighborhood concerns while others were less able to tap into local neighborhood interests. The differences among the leaders' abilities to some extent reflect the types of organizations they represented.

The leaders who were employed by local nonprofit service agencies had an internal support structure consisting of staff and telephones and other office equipment, as well as budget dollars. They were also more likely to have had experience in navigating various bureaucratic channels and in networking to gain funding and support for projects. When these project leaders explained how they went about the business of organizing the planting project, their language showed an awareness of what would be necessary and it reflected this agency-level expertise:

Well first we called a person with the city that gave us some information on who owned the lots and how many were city owned and basically whether we could build on them. So that was one thing to start working on. The second thing was really getting the building demolished and we worked with another person there from the neighborhood city hall office on the east side to just hound the city, to get the building demolished. Then once that was done, then we began to identify our donors and to have the actual [planting] design drawn.

It seems like a simple task to green up a space but in fact it's not. It's something that takes money, dedication, commitment, and then that ability to maintain the property.

Leaders from local neighborhood organizations with a history of community service also had certain advantages. Past experience with successfully completed neighborhood projects gave these individuals credibility with local residents. Their faces were familiar and their methods for organizing neighbors well tested. Unlike the agency-level leaders who merely worked in the neighborhood, these leaders lived in the neighborhood and had a better sense of how to organize neighborhood residents. In addition, their track record of successfully completed projects gave them a sense of competence. Being residents themselves, they had a connection to the area that agency leaders lacked. This was reflected by a passion in their voices when they described local neighborhood concerns:

I started that block club. I lived there so I saw it deteriorating and I started this block club, which took about six meetings before people started coming out. I started knocking on doors and trying to convince the people to come on out, we can change this block.

I believe that grassroots leaders can change their own environment, . . . they could change a lot of things, believe me, cause they live there, they know where the movers and shakers are, you know, where the drug dealers are. You know, we know where the people that will give are. We know that because we live there.

We wanted the vacant lots for beautification and we didn't want the blight there and we didn't want the crime there. So what I did, I talked to the Mayor, the Mayor had a meeting, rather, and I had to stand and I said to him that, concerning the ugly sites. He said "well if the organization can perform the work, I'll give you the opportunity. Send in a written proposal to the DPW." I got together with my recording secretary and we sit down and give it to the Board and they agreed. We sent the letter in with the proposal and it was accepted.

By contrast, leaders from neighborhoods with no history of successfully completed projects, regular meetings, or other group undertakings had more difficulties in organizing their neighborhood tree-planting project. For these individuals, every step taken in organizing the project was a learning experience. Without an internal support structure in place, it was often difficult to overcome a sense of inertia in getting people to participate. In these cases, it seemed that even the connection these leaders shared with their neighbors was not enough to motivate involvement.

[We meet] once a month and we just have a few people coming. If they think there's gonna be some money for the neighborhood, they'll come out in droves, but otherwise just have a few people. We just meet and have a discussion and serve refreshments afterward. Whenever there was discussion they felt that the city was supposed to come in with some money, we had a lot of people, but otherwise they just stay away. It's hard to get people interested.

To a passerby it may appear that a tree-planting project begins and ends on planting day; vital but not visible is the considerable activity that goes on behind the scenes to make it happen. The leaders assume a lion's share of the work and responsibility for making the day a success, yet the approaches they take differ in many respects. These differences are reflected in project outcomes, which can shape how local residents see themselves and their neighborhood. The experience of agency leaders gave them advantages in terms of easily navigating bureaucratic channels or securing funding for the project. Local neighborhood leaders had the advantages of being familiar faces in the neighborhood and knowing local neighborhood concerns. Those with greater organizing experience knew which neighborhood residents to call upon for help. Prior success with other neighborhood projects gave them a foundation from which to conduct additional neighborhood improvement projects.

Follow-up to Planting Day

Seeing that the trees are planted takes substantial effort and makes a perceptible difference at a site. Once planted, however, the trees and the surrounding site need considerable care. Existing grasses and weeds will quickly become overgrown if they are not mowed and if pathways are not mulched regularly. Water must be carried to the site on a regular basis to maintain the trees, shrubs, and flowers planted there. Litter must be picked up. If a neighborhood group cannot give sustained attention to these tasks, the lot will quickly revert to its former condition.

Neighborhood leaders continue to play an important role in ensuring the maintenance of the lots. By signing the maintenance agreement on behalf of the neighborhood organization, their efforts become central to mobilizing resources, motivating local residents, and keeping the lots as positive neighborhood landmarks. Here again leadership differences were quite evident, reflecting how the leaders viewed the project, themselves, and the neighborhood.

The leaders from informally organized neighborhood groups that had participates in few previous group projects voiced frustration and sadness in seeing their good intentions fall short of the mark after planting day was over and the vacant lots returned to their overgrown appearance. A neighborhood leader shared her realization of the amount of effort needed to sustain a lot project:

Well right now it's overgrown with weeds. . . . The big problem is the mowing, the weeds have grown. . . . I would not deal with a vacant lot unless there was a person on that block that was willing to take charge because I mean I live on this street and you know, I'm always mustering people up to do it. I mean I can deal with it now, but I don't know how much longer I can keep doing it.

The situation for leaders with agency support is quite different. Not only do they have a clearer sense of what is involved in maintaining the lots, they have the resources in place to oversee lot maintenance programs. Their projects were maintained either by hired maintenance crews or as part of the agency's program (e.g., a community mental health agency has developed a tree-watering program that is run by their day clients).

Working with the [planting organization], [neighborhood organization], and friends, I've been able to clear that lot, spruce it up, clean it up. We've been able to bring in approximately $3000 worth of trees, shrubs, and bushes that we really wouldn't have been able to afford any other way. . . . We have a crew on a weekly basis maintain that lot and others in northwest Detroit as we work together to stabilize and revitalize Detroit's neighborhoods.

The more experienced neighborhood leaders mobilized local maintenance efforts with ease. Their experience from prior neighborhood projects most likely provided them with knowledge of which residents would take part in maintenance efforts. A reliable means of enlisting neighborhood involvement not only supports ongoing lot care, it sets the stage for residents to see themselves in the activity of tending the lots. The following quotes, from seasoned female neighborhood leaders, provide examples of the investment of local residents in lot care:

As a matter of fact, someone just came and told me yesterday, a guy was doing some work for me here who lives not too far from the [lot], was telling me I need to get somebody over there to cut the grass. I've got to call our president and see if he's found anybody to cut the grass this year.

The children keep it maintained. [A grant] pays the children a volunteer stipend. After they're out of school, all summer, and we use our woodchips. We do still get woodchips from the city.

The last couple years we had the kids do it, we had the kids for they get school supplies. They do it about four times in the summer, pull weeds, you know, clean it up. They do it about three or four times during the summer and they get school supplies for doing it.

While the leaders have a continued role in seeing that planting day is more than a happy moment in the life of the neighborhood, many of them also expressed their appreciation of what the lot keepers accomplish. Not only can we see these individuals as unsung heroes, we see from the leaders' perspective evidence of the identification of workers with the work they perform on behalf of the neighborhood.

Well there's a guy across the street, matter of fact he came here right after I got back and said he "I'm on my way, about 7 o'clock, and the other two guys from the block. I'm on my way down to the park to cut it." You know they just, they just, you know because they feel that [lot] is theirs, because they're the watchers on the block in terms of helping to maintain it and so forth. They look at that as part of their, you know, their job.

Right now the volunteer through [our] greening committee has kind of accepted that [work] and is doing that himself and getting other people involved in caring for [the lot]. . . . Yeah, one person is kind of doing that [maintenance]. He's doing the most difficult task and that's the lawn.

Lot Keepers: The Unsung Heroes

In the months and years after planting day, it is the lot keepers who sustain the lots. As a group, these individuals tended to be less visible than the leaders and often less loquacious. Theirs is not an easy task; many toil for hours on end in their neighborhood lot, often using their own equipment, and with little or no financial compensation.

The thirty-eight lot keepers who participated in the study covered a large range of ages. One maintenance crew consisted entirely of neighborhood youths. Teens were also involved in maintenance work at two other sites. Thus sixteen (42 percent) persons in our sample were under the age of 19. Half that many (21 percent) were individuals over 50, and the remaining 37 percent were between the ages of 20 and 49. Eight of the eleven vacant lot projects were maintained by local neighborhood residents.

Lot Keepers and Expertise

Maintenance work demands a vast array of abilities and knowledge. There are issues of timing, equipment maintenance, nurturing different kinds of plants. Many of those tending the lots had prior experience that enabled their work to progress smoothly. As a group they enjoyed sharing their knowledge:

If a neighborhood or organization wanted to do [a vacant lot project], make sure the commitment is there to follow up through it. Otherwise the whole thing is really a waste of time in the first place. There's gotta be an active commitment to, you know, you can't just do something and leave it and expect it to happen. It won't, you know, and that's what happens time and time in this neighborhood. Someone will get an idea and start something and then it's dead in the water.

You have to give yourself plenty of time. You can't wait until the last minute and say well we'll clean this lot today, because it doesn't work that way. Same applies with mowing of the grass and the upkeep of the grass. You have to go around each and every time and pick up rocks and sticks because they come to the top, the ground swells and the more you water it to keep it nice, to keep the grass growing, the more rocks and sticks come up to the top and then you gotta go around and pick them up. Otherwise you're buying a lawnmower blade or a new motor or something.

Often maintenance work is a way for local residents to apply their interests and talents in service to their neighborhood. Men from three different neighborhoods talk about the technical side of keeping grass and weeds in check:

They need a good lawn mower and make sure the blades are very sharp on it. Cut it [the lot] not too low, say around two and a half, three inches.

That's a John Deere with a 48-inch blade on it, but really you still need larger equipment. You need something with at least a 60-inch cut to handle a lot that size.

See what I do, I have a mulcher, a 5 horse mulcher and I put it on high. I adjust my wheels up, so consequently the grass will grow faster if it's not being cut all the way down, but that way, you know, I don't strain the mower or nothing, you know. And I always just make sure it's good and dry, of course, you know.

Lot Keepers and Identity

Maintaining the lots is much more than the sum of the tasks that need to be completed. For many of the lot keepers their work is a labor of love. Their descriptions of their work include references to the social benefits the neighborhood gains. They see their work on the lots as important, not only in determining who they are, but also as an integral part of shaping the neighborhood scene.

The average block, block club, got a vacant lot. They would want to keep the grass cut. They wouldn't want weeds to grow up on it. I wouldn't want to live next door to a vacant lot, you know I'd be less than a man not to cut it, you know, that's the way I feel.

We even took care of the yard next door. We took care of that, you know, because that building is vacant so nobody's in there and it would make this [lot] look bad. See what I'm saying? If yours is neat and trim and that [one] isn't we would still take care of the yard next door. We try to help the neighborhood look better.

The words of the lot keepers often reflect their strong sense of connection to the local neighborhood and to how their neighbors see and appreciate the work they do. This was true whether one worked as part of the neighborhood or in the context of an agency. What did seem to matter, however, is whether they saw their efforts as successful. If they were experiencing some success in their maintenance efforts, their words underscored the importance of that work to their own neighborhood and to the community in general. The quotes here are from interviews with lot keepers. These four middle-aged men and the young woman are all local residents.

[This is] a big difference to what it was. I mean you'd be surprised. When we're out here cutting [the lot] everyone just slows up and looks . . . and the purpose of that community bulletin board is to get people's attention to what's going on here.

We try to keep it cut, and either end you looking from, it looks nice. And it's really nice too with the trees. I seen a couple people sitting in there one day. That was really nice, you know.

Yeah me and Billie try to be role models for all the other youngsters in the block and try to make them be dependable. Actually me and Billie are the two male role models on the block. I try to be and that give me a lot of inspiration 'cuz a lot of times I be tired 'cuz I try to keep this end [of the block] in order too. So it's rewarding because when you fulfill within yourself, it's rewarding. The park is looking better because the trees are growing. I'd have never dreamed they would be that beautiful.

The neighbors, they appreciate the way we keep the place up. And see, the neighborhood is coming back in terms of the church down there, the great big one down there owns a lot of these homes and they're renovating them. So they're glad because they're putting in new lawns and stuff, so they're glad to see a place like this [lot]. They're glad to see when we put out flowers and stuff, they're glad to see that because that makes the whole neighborhood look better. We try to be good neighbors.

When we were working there [on the lot] a few kids came by and started helping out. And they said well we play here and, you know, things like that. So that was pretty cool. So we actually, we knew why we were doing it, who we were doing it for.

The thirty-eight lot keepers interviewed for this study ranged widely in age, knowledge, and personality. There was the quiet, reserved middle-aged man who took it upon himself to care for a vast stretch of neighborhood vacant lots using his own tools and John Deere riding mower. He did not say much, yet his words reflected an awareness of the way these green spaces connected local residents to the natural world. There were the young charges all dressed in their purple tee shirts who worked with an elderly African-American woman. Their work provided them with something wholesome to do under the direction of a wonderful mentor. There were the two men on the city's east side who took great pride in their work and identified themselves as the male role models on their block. As a group the lot keepers represented such great variety that it is difficult to capture their diverse talents and abilities. Yet their involvement in lot maintenance allowed each of them a context in which to express themselves. Through their efforts, the local stewardship projects flourished, and they in turn found meaning and satisfaction in their contributions.

Community and Identity

The stories told by the leaders and lot keepers tell of complex relationships between personal identity and physical place. The transformation of place, a vacant lot, leads to changes in the neighborhood and the people who live there. Transformation in the people in turn leads to further changes in the neighborhood. To begin the cycle, however, it was necessary for someone to take the initiative. There had to be an individual who had a vision of change; a sense of how to make things happen; and enough patience, persistence, and charisma to lead the way.

In this section we reflect on what we have learned from this and other environmental stewardship programs (Grese, Kaplan, Ryan, & Buxton, 2000; Ryan, 2000; Ryan, Kaplan, & Grese, 2001). In the latter studies, for example, we also found close relationships among transformed places, identity, and involvement. In some instances, the decision to participate in activities such as removal of invasive plant species or river monitoring is based on the volunteer's conviction about helping the environment. In other instances, it is through increasing experience that stewardship volunteers discover a sense of purpose and clarity that provides new directions and identity. In the context of the vacant lot projects, it is too soon to know whether for the younger lot keepers their involvement, which generally began as happenstance, might turn into life-transforming missions.

Transformation of Place and Person

The process of greening a local vacant lot is cyclical in nature. Just as the project changes over time, so too do those who tend it and use it. The lots provide physical examples of the interplay among involvement, expertise, and identity. The circumstances under which one chooses to participate in a project and the outcomes of that participation can affect both continued involvement and how individuals feel about the work they do.

In some neighborhoods the vacant lot project is one of the first projects undertaken, so the cyclical pattern is less evident. It is easier to see the cycle in neighborhoods where the vacant lot project is one in a long line of successful projects that have shaped present-day neighborhood identity. This is nicely illustrated by the following quote from a neighborhood leader who has spent more than a decade organizing projects in her community:

I think the first project that we worked on was the lawn lights. Everybody wanted one of those lawn lights. I had someone come out from one of the companies and talk about the lawn lights at the next meeting. That really started bringing people together. The day they installed those lawn lights was the day I knew there could be unity on that block because everybody started coming out and looking at them installing lights. 'I'm gonna put bricks around mine.' 'I'm gonna put flowers around mine, and gravels here.' It really brought the people together. . . . And you know signs went up, and we started painting the curbs. Everybody started noticing that one block.

The cyclical nature involves other facets as well. Pride in the neighborhood increases, along with self-esteem. People become more willing to participate, leading to changes even beyond the vacant lot. It is characteristic of many environmental stewardship projects that participation in the tangible changes to the setting—tree planting, removal of invasive species, river cleanups—is accompanied by personal changes.

The underlying relationships among these qualities are difficult to document. However, it is less difficult to believe that these transformations are happening when one hears local residents reflect on the changes they have witnessed. One of the agency leaders clearly understood these relationships:

So the neighborhood, what we've been trying to do with the whole neighborhood is to kind of lift it up and to improve it and to rebuild and to be a part of the rebuilding that's coming this whole way. 'Cause it just makes a difference, your environment, how you act and how you feel about yourself.

Leadership-Driven Outcomes

Vacant lot planting projects involve more than trees, shrubs, and flowers. They are part of a local neighborhood social scene and are shaped by the efforts of many individuals. The methods used for planning, planting, and maintaining these projects can have profound consequences for the local community. Thus the expertise of the leaders and their approaches to implementing the project have substantial impact on community identity.

Leaders from agency-led projects make good use of organizational skills and support in carrying out their projects and attending to follow-up maintenance. This agency expertise is quite successful at transforming weedy, overgrown lots into well-kept neighborhood parks and green spaces. At the same time, however, the lack of a local neighborhood presence or connection to local neighborhood concerns lessens the opportunities for involvement by local citizens. This lack of connection to surrounding neighborhood residents can mean missed opportunities for bolstering community identity as the project evolves.

By contrast, leaders from locally established neighborhood organizations, using tried-and-true methods of community organizing and representation, set the stage for neighborhood involvement in these projects before, during, and after planting day. Planned with an eye toward neighborhood wishes and uses, these projects are more inclined to become highly valued community spaces. Involvement by local residents, particularly in lot maintenance, further increases opportunities for weaving the project into the existing social identity of the neighborhood.

Reliance on local leadership, however, provides no assurance of positive outcomes. A local leadership that lacks strong organization and a history of successfully completed projects may ultimately defeat the project. The highly visible nature of these projects, both before and after planting day, broadcasts failure that extends beyond the loss of trees (see chapter 9).

An Invisible Process

Many aspects of the failure or success of these projects, including the multitude of steps needed to make a project happen and the sustained effort required to maintain it, are hidden from view. Decisions to undertake planting projects may be made without knowing the demands that they will entail. Yet success, with all its ramifications for the neighborhood, often depends on many unforeseen factors.

If it takes so much skill and effort to nurture a planting project, ideally the individuals with experience would share their knowledge and provide assistance to others. Unfortunately this is rarely the case because leaders and lot keepers tend to underestimate their role and understanding; they may feel that they contributed little, that anyone could have done it. This pattern is characteristic of many experts. In gaining expertise, involvement and experience change how we see things (Kaplan and Kaplan, 1982). People tend to be unaware of these subtle but powerful changes; the new way of seeing seems so obvious. Thus while leaders may recognize activities that require support from governmental agencies and those that will gain momentum through informal networking, they may not recognize that this takes special knowledge. Similarly, experienced lot keepers can anticipate how much effort is needed to maintain a project without appreciating that being able to do this is indicative of their expertise.

The invisibility of expertise to the expert thus leads to the tendency to take one's knowledge for granted and makes it more difficult to transfer it to others. The leaders' and lot keepers' expertise is essential to their own identity and to the success of the project; the failure to recognize it, however, can mislead others into thinking that a tree-planting project is a simple and straightforward solution to neighborhood malaise.

Success and Identity
Success has many positive implications, but projects are not always successful. Failed projects are not only damaging to the neighborhood, they are deeply painful to those involved. We heard it in the voices of leaders whose visions for the project did not match the outcomes. We also heard it as lot keepers described the futility of trying to reclaim a vacant lot that all too quickly became overgrown and unattractive. The process of trying and then failing, especially when repeated again and again in a neighborhood, can have profound impacts upon how those involved come to think of themselves and their neighborhood. It is not surprising that local residents are often reluctant to become involved in projects like these, particularly when they are undertaken by outside agencies or occur in neighborhoods that have experienced few successes.

The lesson here goes beyond the impacts of a failed project upon local neighborhood residents. There are risks and dangers in carrying out a project without careful consideration of how it will be tended after planting day or how it will be woven into the fabric of a local neighborhood. To focus solely on projects that work or are successful is as dangerous as it is irresponsible. The powerful effect of a project that falls short of the mark provides ample evidence of why these ventures should not be undertaken lightly.

With sufficient examples of failed efforts around the city, it is all the more remarkable to see so many instances of planting projects that have many of the marks of success. Some of these once-vacant lots have become important foci for the neighborhood. They are places residents seek for quiet moments, where they know they will meet others or learn about local happenings, or where local events take place. The residents see local, familiar faces tending and using the lots, and this reinforces their belief that the lot belongs to the neighborhood.

Concluding Thoughts

The process of transforming vacant lots into neighborhood oases appears to have important psychological benefits. By contrast, failed efforts to create a new and valued neighborhood place may unfortunately undermine both personal and community identity. This study suggests a number of hypotheses concerning the roles these places play when the attempted transformation is successful.

The Lot as Community Landmark

The open space itself provides a visual resource that is a frequently encountered part of the local scene. As residents see what is happening on the lots and watch them thrive under local care, they come to consider them a vital part of the neighborhood. The lots thus become more than a place to notice; they become locally significant places to share and use. The presence of an attractive, shared, and highly visible location is hypothesized to enhance the community's sense of identity.

Linking Involvement and Identity

Personal identity is intimately related to the place where one lives and how its transformation is perceived by the self and others. This is augmented by the degree to which one has been personally active and involved in the transformation. Participation in creating the place or in sustaining it as a local resource is hypothesized to enhance personal identity through an increased sense of connection and ownership, as well as through receiving the respect of the community.

Toward Continued Changes

The involvement itself is also a source of psychological benefits that are self-supporting. Participation is hypothesized to promote further involvement that ramifies and extends to other projects, further enriching the neighborhood and possibly encouraging others to participate in the process. Involvement, expertise, and identity are thus closely intertwined. The success of tree-planting programs depends on them and helps foster them as well.

Acknowledgments

We wish to thank the Detroit residents who participated in these interviews; their insights and knowledge are invaluable. We also are grateful for the support of USDA Forest Service, North Central Forest Experiment Station through several cooperative agreements. Our thanks to Susan Clayton and Susan Opotow for their thoughtful review and helpful comments, and to Stephen Kaplan for valuable feedback throughout the writing process.

References

Austin, M. E. (1999). Dimensions of resident involvement generated by inner-city tree planting projects. Doctoral dissertation, University of Michigan, Ann Arbor.

Darden, J. T., Hill, R. C., Thomas, J., & Thomas, R. (1987). *Detroit: Race and uneven development*. Philadelphia: Temple University Press.

Grese, R. E., Kaplan, R., Ryan, R. L., & Buxton, J. (2000). Psychological benefits of volunteering in stewardship programs. In P. H. Gobster & R. B. Hull (Eds.), *Restoring nature: Perspectives from the social sciences and humanities* (pp. 265–280). Washington, D.C.: Island Press.

Grove, M., Vachta, K. E., McDonough, M. H., & Burch, Jr., W. R. (1993). The urban resources initiative: Community benefits from forestry. In P. H. Gobster (Ed.), *Managing urban and high-use recreation settings* (pp. 24–30). General technical report NC-163. USDA Forest Service, North Central Forest Experiment Station, St. Paul, Minn.

Kaplan, S., & Kaplan, R. (1982). *Cognition and environment: Functioning in an uncertain world*. New York: Praeger. (Republished 1989 by Ulrich's Ann Arbor, Mich.)

Ryan, R. L. (2000). A people-centered approach to designing and managing restoration projects: Insights from understanding attachment to urban natural areas. In P. H. Gobster & R. B. Hull (Eds.), *Restoring nature: Perspectives from the social sciences and humanities* (pp. 209–228). Washington, D.C.: Island Press.

Ryan, R. L., Kaplan, R., & Grese, R. E. (2001). Predicting volunteer commitment in environmental stewardship programmes. *Journal of Environmental Planning and Management, 44*(5), 629–648.

11

Representations of the Local Environment as Threatened by Global Climate Change: Toward a Contextualized Analysis of Environmental Identity in a Coastal Area

Volker Linneweber, Gerhard Hartmuth, and Immo Fritsche

Coastal areas offer people various elements of nature: land, sea, wind, and their interplay. Throughout the world, people prefer living in coastal areas (Geipel, 2001) and like spending leisure time there, watching the surf and taking photographs of spectacular shorelines. Real estate near the sea is attractive and expensive. In the past, sea-based transportation and the availability of seafood contributed to the attractiveness of coastal areas, but the importance of these factors has diminished with the rise of modern transportation methods, so this does not fully explain people's sustained preference for coastal areas; something else must be the cause of this fascination. We study environmental identity in a specific coastal locale—the North Sea island of Sylt—to understand how people define themselves, others, nature, and environment in one specific region.

Our starting point is that nature and environment have meaning and significance for people. Stern, Young, and Druckman (1992, p. 34) use the phrase "what humans value" to address the personal significance of environmental states and changes. Although we might assume that a farmer has a different relationship with nature than a stockbroker, the farmer might spend his leisure time trading stocks while the stockbroker might enjoy gardening. Thus the personal meaning of nature for both might be less stereotypical than we suppose.

Although systematic research is not yet available, we suspect that the natural environment is more significant for people living in coastal areas than for those dwelling inland. One cause of this is their immediate exposure to nature and its dynamic elements. In addition to its personal significance, residents of coastal environments observe that their environment is attractive for others. Thus coastal dwelling has social meaning as well as well as personal and private meaning.

Coastal areas are not only highly attractive, but they are also vulnerable to environmental impacts. This impact varies as a result of land use and geographic location. This vulnerability may be an additional element of the attractiveness of coastal areas. Environmental threats to such areas, which include hydrodynamic impacts, the possible rise of sea level as a result of climate change, and the effects of intense storms increase the importance of nature and the environment for the residents affected. In less exposed areas, residents may occasionally spend time thinking about nature and the environment, but in coastal areas the natural environment commands more attention. Vulnerability and the means to cope with it are subject to societal discourse. This includes debate about how coastal areas ought to be utilized, how one evaluates vulnerability, and how individuals and communities can cope with hazardous environmental change. We suspect that coping with environmental threats and hazards and witnessing environmental impacts is an important aspect of environmental identity in coastal regions.

Using a case study on social representations of global climate change in a coastal area, we argue that the concept of environmental identity is part of the more general concept of situated identity, composed of various facets and subject to social mediation. We first outline our understanding of environmental identity, highlighting the important role of context. Then we describe objectives, methods, and results of a case study on the island of Sylt. We close with a discussion of environmental identity in the light of these data.

Environmental Identity

We define environmental identity as the sum of a person's perceptions and evaluations concerning nature and environment and their relevance for respective everyday contexts. In contrast to well-known conceptual deficiencies in attitude research and studies on environmental awareness that, through prompting effects and social desirability, make "environment" or "nature" salient (e.g., in questionnaire items), we argue that environmental identity has to be conceptualized with respect to everyday situations.

William James (1890) argued that people do not have a single self-concept or social identity, but instead have many different social selves.

Modern research on self-concept and identity incorporates this position. From comprehensive research on personal identity, studies on self-concepts, work on situated identity (Alexander, Knight, & Knight, 1971) and, particularly, social identity theory (Tajfel, 1978), this integration has been achieved. According to Tajfel (1978) (see later studies within this framework), situations vary along a dimension of people acting on the basis of personal or social identity. Extending this position, we postulate that situations also vary according to the importance of additional facets of identity. Environmental identity is one of these additional facets. Therefore, environmental identity is one facet of a situated or contextualized identity. Environmental identity is not absolute, but instead is a relative concept, since valuing nature or the environment competes with additional facets of identity (Bossel, 1990). This is illustrated by the conflict between economy versus ecology in the discussions of general policy goals.

The idea of attachment to place as a component of identity is rooted in the social psychological and environment and behavior literature. Proshansky and colleagues (Proshansky, 1978; Proshansky, Fabian, & Kaminoff, 1983) and later Stokols and Shumaker (1981) initiated this line of research, considering "place identity" as an integral part of the self and connecting place with the psychology of personality. Johnson (2001) has stated: "The most agreed upon definition of [place] attachment is that of a deep, positive, affective bond to a setting or type of setting. This bond has less to do with rational thought, as in the case of establishing satisfaction . . . rather, it is determined more by emotion in that attachments may be formed with objects or places which are undesirable to the objective observer" (p. 5).

Most studies focus on urban design (Hillier, 2002); communities and specific groups, such as children (Spencer & Woolley, 2000) and elderly people (Norris Baker & Scheidt, 1990; Selby, Anthony, Choi, & Orland, 1990); or specific aspects, such as neighborhood ties (Christakopoulou, Dawson, & Gari, 2001; Mesch & Manor, 1998; Perkins & Long, 2002). Only a few studies focus on place attachment and wilderness or nature (Glassman, 1995; Ryan, 1998; Steel, 1995; Williams, Patterson, Roggenbuck, & Watson, 1992).

Theoretical considerations of concepts like place identity and place attachment have been influential in our conception of environmental identity, but our focus is not the same in all aspects. Considerations of use history indicate that the place attachment concept primarily relates specific places to specific individuals. Deviating from this, our conceptualization elaborates the relationships between types of places and types of actors. The focus is not on an individual use history of a specific place, but on individual and/or superindividual experiences with types of places. Also, our conceptualization does not postulate a dominance of nonrational aspects (Johnson, 2001). The actors responsible for coastal zone management may have a highly rational relation to their field of professional activity. Nevertheless we consider their perception and assessment concerning the impact of high tidal waves as being part of their (professional) environmental identity resulting from their attachment to (types of) places.

We argue that environmental identity is not at all arbitrary but is systematically related to objective interrelations between social participants and environmental entities. As previously indicated, farmers and brokers may differ in their interdependence with nature and environment. Although it refers to intrapersonal processes, environmental identity is subject to social influences. The conceptualization outlined here expands narrower definitions of environmental identity, such as urban identity (Lalli, 1992) or place identity (Korpela, 1989). It attempts to incorporate multiple facets of the environment on a spatial scale: local, regional, and global. As with place identity or urban identity, our conceptualization of environmental identity is not restricted to cognitive aspects but also integrates emotional facets such as attachment to environmentally defined entities.

For decades, the societal discourse on environmental problems has contributed to a perception of nature and environment as—at least partly—unstable, fragile, and risky for humans utilizing it. People now increasingly take into account both environmental benefits and environmental threats. These aspects are incorporated in our conceptualization of environmental identity. Our investigation of environmental dynamics includes people's implicit conceptualizations of environmental risks.

Significance of Environmental Identity

The relationship between environmental attitudes and behavior, like that between attitudes and behavior in general (Lévy-Leboyer, Bonnes, Chase,

Ferreira-Marques, & Pawlik, 1996), is not strong. However, environmental identity has behavioral consequences, like other facets of identity. As a concept with some intraindividual consistency, it influences decisions and personal behavior. Examined across individuals, environmental identities contribute to explaining the dynamics of social systems concerning nature and environment.

The conceptual position sketched out here does not imply that environment-related decisions and behavior are exclusively driven by environmental identity. Lantermann (1999) states that behavior protecting the environment is polytelic and shows that several driving forces may determine similar behavior. This position is compatible with sociological research indicating that different life-styles may result in similar patterns of resource consumption although the motives may differ significantly (Reusswig, 1994). What does this mean for the relationship between environmental identity and environmentally relevant decisions and behavior? First, we should not assume that all behavior directed toward the environment results from environmental identity. Second, we cannot interpret environmentally relevant behavior as an unequivocal indicator for environmental identity.

Environmental identity is not an attribute that can be observed directly. Instead, a variety of indicators must be utilized. Our case study uses two types of indicators: (1) the relative salience of individual representations concerning the environment in the context of other locally relevant states and processes and (2) people's environmentally relevant positions in the social system under investigation as indicated by profession or political involvement. We are also interested in the interrelations among the components of these two indicators.

In connection with the first indicator, we are interested in the relationship between proximal (i.e., local, regional) and distal (i.e., global) environmental aspects. It is important to examine a spatial dimension that ranges from local to global. During the past five decades the societal discourse on environmental problems has shifted gradually from proximal topics (pollution of streams and rivers) to distal (acid rain) and entirely global ones (greenhouse effects). However, if regional or local impacts were not a source of concern, climate change might attract less societal interest and subsequently stimulate less scientific research. Hence, in Stern et al. (1992), the explicit focus on "what humans value" directs our attention toward people's perceptions and evaluations of environmental change.

However, the relationship between global climate change and regional or local impacts is neither clear nor unambiguous. This is even more true for perceived impacts. Pawlik (1991) has described psychologically relevant characteristics of climate change that impede an accurate perception from an individual's point of view, and Linneweber (1995, 1999) has added considerations from social and environmental psychology. Bell (1989, 1994) and others (e.g., Bostrom, Morgan, Fischhoff, & Read, 1994; Kempton, 1991) have demonstrated that information on climate change and global environmental change is confused in the public's understanding. To understand environmental identity, the interrelations between these topics is of great interest since psychologically relevant variables such as coping potential or locus of control vary among people (Pawlik, 1991). One can assume that proximal environmental issues are more influential and consequently contribute more to environmental identity than distal ones. However, a salient environmental identity may mediate the relationship between remote global effects and their perceived personal relevance.

The Island of Sylt: A Case Study

The natural sciences have advanced our understanding of climatology, but in recent years a core question has emerged: How can we explain the functioning of social systems facing global climate change? To answer this question, we need to know if and how potential climate change is anticipated, perceived, and judged by individuals and groups. A promising approach to this question is to utilize case studies in small, manageable social systems that are exposed to climate change. As part of an interdisciplinary research project on climate change and coastal protection on the German island of Sylt (Daschkeit & Schottes, 2002), we are studying social representations of threats to the island, which faces possible impacts of climate change (Linneweber, Hartmuth, Deising, & Fritsche, 2001).

Sylt is a narrow, 40-km-long island of about 100 sq km, with a population of 20,000, located in the German part of the North Sea. The island was selected for study because of its exposure to storms that are increasing in intensity and frequency as well as to hydrodynamic effects that are most likely amplified by climate change. As figure 11.1 indicates, the shape of the island, its function as a natural breakwater, its sandy coastline, and the continuous change in its shape explain the attention to climate change impacts in the local media.

Attempts to stabilize Sylt go back to about 1850. Since its relatively natural coastline contributes to the island's identity and image, in the past 30 years, at a cost of $5–10 million/a year, these efforts have included continuously replacing lost sediment and strict rules that protect its dunes. Because of its coastline and natural sandy beach, Sylt is very attractive to tourists as well as permanent residents, particularly wealthy and elderly persons. Now, however, climate change may increase environmental impacts that threaten the island. Storms and high-tide events are expected to increase in magnitude and frequency (Berz, 2001).

We have examined local groups' understanding of these developments on the island, including events that take place in the natural environment. We have also analyzed their understanding of the causes, effects, and feasible countermeasures for climate change (see Linneweber et al., 2001). These contextualized perceptions and evaluations, which are sensitive to intergroup differences, allow us to understand the subjects' environmental identity. In contrast to a research approach that separates environment-related matters from other topics, our study interprets an understanding of environmental identity in relation to such other aspects of identity as economic and societal identity. Despite the unique set of situational constraints, the results of this case study may apply to other areas that are at risk as a result of climate change, such as mountainous regions in danger of increasing mudslides. The selection of this high-liability area is also intended to provide lessons for more moderate environmental conditions.

Objectives and Hypotheses

The perception and evaluation of climate change have been investigated in various studies (e.g., Bell, 1989; Bostrom et al., 1994; Dunlap, 1996; Henderson-Sellers, 1990; Kempton, 1991; McDaniels, Axelrod, & Slovic, 1996; Read, Bostrom, Morgan, Fischhoff, & Smuts, 1994). Most of these studies, however, focus directly on mental images of the climate problem. Contextual matters as well as more distant causes and effects

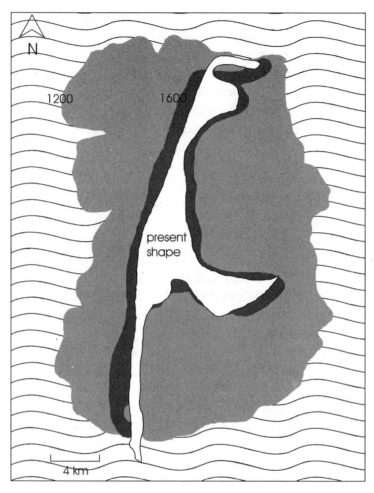

Figure 11.1
The island of Sylt: change of size and location from 1200 A.D. to the present.

have not yet been examined. In our study, we explore subjective under-standings of climate change, looking at the relative importance of these understandings within the context of other locally relevant issues. We hypothesize that these contextualized understandings serve as indicators of environmental identity. Because Sylt is actually threatened by climate change and its impacts, we expected that the topic itself and other topics associated with it (e.g., coastal protection) would play an important role in the understanding of the inhabitants surveyed.

As a second objective, we looked for differing understandings among different groups in the social system on Sylt. One fundamental finding of attitude research is that people tend to keep attitudes and opinions free of contradictions (Festinger, 1957). Incorporating this proposition into the sociopsychological concept of social representations (Moscovici, 1976; 1984; Moscovici & Duveen, 2001), we expected that different groups participating in the social system on Sylt would have different interests and therefore also differ in their shared understanding of general developments. In line with our conceptualization of environ-mental identity, the position of a participant within the social system has to be considered as significantly influencing his or her environmental identity.

Method

Methodologically, research on environmental perceptions and evalua-tions—and hence on environmental identity—is difficult. Because pro-tecting nature and preserving the environment are values with high social desirability, the case of random sample surveys may be insufficient. Therefore, instead of a representative sample, a relatively small number ($N = 70$) of local key persons were surveyed in our study. In line with similar research (Dörner, Hofinger, & Tisdale, 1999; Dörner, Kruse, & Lantermann, 1992) we assumed that these key persons—systematically chosen as members of participating groups—should indicate the social representations of region-specific understandings more accurately than a random sample of subjects. The configuration of our sample reflects the social system of Sylt in that all participating groups of contextual inter-est were represented. Therefore, we proceeded in a pragmatic, criteria-oriented way in reconstructing this social system (see Linneweber et al., 2001 for details).

An extensive analysis of various information sources concerning Sylt led to the identification of roughly 200 key persons, of whom 70 were included in the final sample. The average age of the subjects was 53 years (median), the youngest being 28, the oldest 85. Twenty-three percent of the participants were female, 77 percent male. The level of education in the sample was exceptionally high. Each subject was assigned to one of nine groups: politics and administration (12), tourism (8), trade and industry (except tourism; 12), conservation and environmental protection (7), coastal protection (7), press and media (6), art and culture (6), and education and social work (8). Four additional participants from outside Sylt were involved as decision makers.

Semistructured face-to-face interviews were conducted with each subject in March 1998. The subjects were asked about their perceptions and opinions concerning general developments on the island in the three broad areas of economy, society, and natural environment. All sixty-nine interviews (there was one failure that was due to technical reasons) were taperecorded and transcribed. An extensive category system was developed in order to have the interview protocols systematically coded by trained student assistants. The reliability of the codings was controlled by calculating the coefficients of interrater agreement (Cohen's kappa) for randomly selected interviews and reached an average of $\kappa = 0.82$. The eighty-five specific coding categories were aggregated to fourteen more general topics. These were analyzed by quantitative content analysis, based on the frequency of their occurrence in complete answers to interview questions. Since answers to eight interview questions were considered, a maximum of eight references to one topic were included in the counting of frequency per person, absolute counts being divided by the total number of answers analyzed (541). In contrast, differences among participating groups were analyzed on the basis of the 5,561 original codings.

Results

In the interviews, the subjects were asked to describe past and future developments concerning Sylt from their perspective. Input from the interviewer was restricted to addressing three main areas—the economy, society, and nature and environment—by general questions. Thus statements indicating environmental identity are framed by perceptions and evaluations of other aspects.

Table 11.1 lists the respective percentages of answers in which each of the fourteen topics were mentioned. (These topics are italicized in this text.) *Tourism* and *building and land use* were of outstanding importance in the subjects' understandings of general developments concerning Sylt. This result is due to the economic significance of tourism for the island as well as its obvious side effects. Apart from this, it is quite remarkable that aspects summed up in the topics of *nature and environment* (e.g., nature and landscape as the economic basis of the island), and, to a somewhat lesser extent, *conservation and environmental protection* (e.g., measures to protect nature and the environment) follow next in the table, ahead of other topics. The topics of *coastal protection* (e.g., measures to stabilize the coastline) and *changes in the shape of the island* (e.g., reduction of the island in size and substance), however, which are highly relevant in the context of local impacts of climate change, were mentioned in considerably fewer answers.

Table 11.1
General developments on the island of Sylt

Topic	%
Tourism	57
Building and land use	54
Nature and environment	33
Conservation and environmental protection	24
Living	23
Politics	22
Demography	22
Quality of life	21
Other economic areas (than tourism)	21
Mobility and traffic	20
Coastal protection	16
Culture	15
Changes in the shape of the island	14
Labor and employment	13

Note: Relative frequencies are in percentages, based on all answers given in the interview.

Taking the relative weight of respective topics as an indicator for environmental identity, one can conclude that nature and environment contribute moderately to environmental identity, whereas the potential threat to the island from global climate change plays a subordinate role. This can also be observed when analyzing the coded categories at the nonaggregated level. For example, the specific category *climate change* itself was mentioned in only 4 percent of all answers.

To discern different patterns in the subjects' representations according to their position in the social system, we analyzed the frequencies of those topics closely corresponding with the three most distinctive participating groups ("conservation and environmental protection," "coastal protection," and "tourism"), separately comparing each group with the pooled rest of the sample. We expected that the subjects would prioritize topics that could be considered as defining their group, mentioning those aspects more frequently than all other subjects. The results are shown in table 11.2.

Table 11.2
General developments on Sylt (selected topics): Differences among particular participating groups and all other subjects, respectively

Participating group and topic	Frequencies (%)[a]		$\chi^2_{(df=1)}$	p
	Group	Others[b]		
Conservation and environmental protection group				
Nature and environment	5	6	0.26	0.61
Conservation and environmental protection	9	5	15.02	<0.001
Coastal protection group				
Changes in the shape of the island	4	2	13.30	<0.001
Coastal protection	8	3	51.99	<0.001
Tourism group				
Tourism	22	14	37.13	<0.001

[a] Relative frequencies in percentages, based on all codings.
[b] Pooled group of all other subjects except respective participating group.

Consistent with our hypothesis, members of the participating group "tourism" mentioned the topic of *tourism* more frequently than other participants. The same holds for the participating group "coastal protection" and the respective topics, although on a lower level of mention. A distinction also emerged between "response" topics (those involving an active human response) and "state" topics (those referring to some state or process of nature). There was a bigger effect for the response topic *coastal protection* (e.g., taking measures to stabilize the coastline) compared with the state topic *changes in the shape of the island* (e.g., reduction in the island's size). As for the participating group "conservation and environmental protection," group affiliation produced a significant difference only for the response topic, *conservation and environmental protection* (e.g., taking measures to protect the environment), whereas the state topic, *nature and environment* (e.g., the economic functions of nature and landscape), was mentioned by both groups at about the same frequency. Although further substantiation of these results is required because of the small frequencies, we interpret them as consistent with our assumption of group specificity for particular topics.

What makes the results noteworthy with respect to environmental identity (utilizing the concept in the broad meaning of situated identity here) is first, the fact that the topic of *tourism* was mentioned by the participating group "tourism" far more frequently than the other topics under consideration by the respective groups. So tourism might have a higher impact on the identity of the subjects engaged in this field than topics in the nature and environment field for the respective groups. Second, aspects of the state of nature and the environment were shared by all the subjects surveyed; they were not at all specific for the respective participating group. So nature and environment—apart from conservation measures—seem to be part of the identity of the whole sample.

Discussion

There are complex person–environment interrelations in this coastal area facing the effects of global climate change. Our results elucidate the concept of environmental identity in ways that are compatible with the more general concept of situated identity. At first glance, these data indicate that within a sample of key persons on Sylt, nature and environment in general receive only moderate consideration, indicating a low level of environmental identity. Instead, the development of tourism as the main source of revenue as well as the use of the island's expanse of ground were topics with which the subjects were more concerned. However, these results are more complex. Some topics that are related to environmental identity are well represented in the findings: nature and environment, conservation, and environmental protection. Apparently, general aspects of the physical environment and its conservation are seen as important topics with respect to continuing development on the island. Perhaps this reflects the symbolic and economic value of the island's natural environment. Changes in the shape of the island and coastal protection, environmental topics that are closely associated with the impacts of climate change, are represented to a lesser degree, in contrast to our hypothesis.

The topics mentioned in the interviews could imply that cognitions relevant to climate change are not important for this sample's environmental identity. However, a comparison with the results of a second set of interviews dealing with climate change topics in detail, and of an additional "quick-alert-study" triggered by an actual hurricane, leaves a different impression. Most participants showed a detailed knowledge about the climate change problem, attributed a high likelihood to its occurrence, and also declared that it was responsible for the hurricane's intensity. This pattern of results, which are currently being analyzed in detail, leads to the assumption that the participants of our study did not link their knowledge about *global* phenomena to the cognitions that form their *local* environmental identity.

Concerning differences among participating groups, our empirical evidence supports our hypothesis that social representations will vary with the position of participants in the social system. In contrast to this, however, all the subjects gave a comparable amount of consideration to the state of nature and the environment as a whole. Both results support our position that a contextualized analysis of environmental identity extends the concept. The more general notion of situated identity may still be described as environmental identity, but it utilizes a very broad meaning of "environment" that includes the whole local situation. In such a concept, "nature" and "environment" are facets with only relative significance among other situational topics.

The notion of contextualization is central to our conception of environmental or rather situated identity. The term *contextualization* has two meanings here. On the one hand, it means that topics not directly associated with nature and the environment (e.g., tourism in our study) should be analyzed as frames that indicate the relative importance of "truly" environmental topics in creating an environmental identity. On the other hand, it means that at least parts of an environmental identity are socially mediated, with different participating groups showing different identities according to their position within the social system.

Supported by attitude research telling us to carefully monitor the effects of social desirability, we suggest studying social representations of logically—but not necessarily psychologically—related environmental processes in an ecologically valid way (Brunswik, 1956). This enables us to elaborate the relevance of environmental identity for everyday decisions and behavior and also complements studies on mental models of climate change. Even if the "key person" approach may have some shortcomings (e.g., regarding the diversity of the sample under study). We suggest studying environmental identity in environmentally extreme contexts, relating it to the environmentally relevant positions of participants within the local social system, and associating local with global aspects of environmental issues. It is our hope that this approach will stimulate further research, which is definitely needed.

Acknowledgment

The case study reported in this chapter was supported by a grant from the German Federal Ministry of Education and Research (01 LK 9563).

References

Alexander, C. N., Knight, G. W., & Knight, J. R. (1971). Situated identities and social psychological experimentation. *Sociometry, 34,* 65–82.

Bell, A. (1989). Hot news: Media reporting and public understanding of the climate change issue in New Zealand. A study in the (mis)communication of science. Department of Linguistics, Victoria University, Wellington, New Zealand.

Bell, A. (1994). Climate of opinion: Public and media discourse on the global environment. *Discourse & Society*, 5, 33–64.

Berz, G. (2001). Naturkatastrophen an der wende zum 21. Jahrhundert: Weltweite trends und Schadenspotentiale [Natural disasters in the beginning of the 21st century: Worldwide trends and damage potential]. In V. Linneweber (Ed.), *Zukünftige Bedrohungen durch (anthropogene) Naturkatastrophen* [Future threats by (anthropogenic) natural disasters] (pp. 4–14). Bonn: Deutsches Komitee für Katastrophenvorsorge.

Bossel, H. (1990). *Umweltwissen: Daten, fakten, zusammenhänge* [Environmental knowledge: Data, facts, interrelations]. Berlin: Springer-Verlag.

Bostrom, A., Morgan, M. G., Fischhoff, B., & Read, D. (1994). What do people know about global climate change? I. Mental models. *Risk Analysis*, 14(6), 959–970.

Brunswik, E. (1956). *Perception and the representative design of experiments*. Berkeley, Calif.: University of California Press.

Christakopoulou, S., Dawson, J., & Gari, A. (2001). The community well-being questionnaire: Theoretical context and initial assessment of its reliability and validity. *Social Indicators Research*, 56(3), 321–351.

Daschkeit, A., & Schottes, P. (Eds.). (2002). *Sylt. Klimafolgen für Mensch und Küste* [Sylt: Climate impacts for society and coastal zones]. Berlin: Springer-Verlag.

Dörner, D., Hofinger, T., & Tisdale, T. (1999). Forschungsvorhaben "umweltbewusstsein, umwelthandeln, werte und wertewandel." Endbericht [Research project "Environmental concern, behavior, values and value change." Final report.]. Institute for Theoretical Psychology, Otto Friedrich University, Bamberg, Germany.

Dörner, D., Kruse, L., & Lantermann, E.-D. (1992). Umweltbewusstsein, Umwelthandeln, Werte, Wertwandel. Antrag auf Förderung eines Forschungsvorhabens [Environmental concern, behavior, values, value change. Research Proposal.]. Unpublished manuscript.

Dunlap, R. E. (1996). Public perceptions of global warming: A cross-national comparison. In Human Dimensions of Global Environmental Change Programme (Ed.), *Global change, local challenge. HDP Third Scientific Symposium, Vol. 2. Poster papers* (pp. 121–138). Geneva: Human Dimensions of Global Environmental Change Program.

Festinger, L. (1957). *A theory of cognitive dissonance*. Stanford, Calif.: Stanford University Press.

Geipel, R. (2001). Zukünftige Naturrisiken in ihrem sozialen Umfeld [Future natural risks in their societal context]. In V. Linneweber (Ed.), Zukünftige Bedrohungen durch (anthropogene) Naturkatastrophen [Future threats by (anthropogenic) natural disasters] (pp. 29–38). Bonn: Deutsches Komitee für Katastrophenvorsorge.

Glassman, S. B. (1995). The experience of women discovering wilderness as psychologically healing. *Dissertation Abstracts International*, 56(04B), 2325.

Henderson-Sellers, A. (1990). Australian public perception of the greenhouse issue. *Climatic Change, 17,* 69–96.

Hillier, J. (2002). Presumptive planning: From urban design to community creation in one move? In A. T. Fisher & C. C. Sonn (Eds.), *Psychological sense of community: Research, applications, and implications. The Plenum series in social/clinical psychology* (pp. 43–67). New York: Kluwer Academic/Plenum.

James, W. (1890). *Principles of psychology* (Vol. 1). New York: Henry Holt.

Johnson, C. Y. (2001). A consideration of collective memory in African American attachment to wildland recreation places. Northwest Minnesota Foundation. Available at http://www.srs.fs.fed.us/pubs/rpc/2000–05/rpc_00may_38.pdf [2002, 10–31].

Kempton, W. (1991). Lay perceptions of global climate change. *Global Environmental Change, 1,* 183–208.

Korpela, K. M. (1989). Place identity as a product of environmental self-regulation. *Journal of Environmental Psychology, 9,* 241–256.

Lalli, M. (1992). Urban-related identity: Theory, measurement, and empirical findings. *Journal of Environmental Psychology, 12,* 285–303.

Lantermann, E.-D. (1999). Zur Polytelie umweltschonenden Handelns [The polytely of ecologically oriented behavior]. In V. Linneweber & E. Kals (Eds.), *Umweltgerechtes handeln: Barrieren und brücken* [Ecological behavior: Barriers and bridges]. Heidelberg: Springer-Verlag.

Lévy-Leboyer, C., Bonnes, M., Chase, J., Ferreira-Marques, J., & Pawlik, K. (1996). Determinants of proenvironmental behaviors: A five-countries comparison. *European Psychologist, 1,* 123–129.

Linneweber, V. (1995). Evaluating the use of global commons: Lessons from research on social judgment. In A. Katama (Ed.), *Equity and social considerations related to climate change* (pp. 75–83). Nairobi: ICIPE Science Press.

Linneweber, V. (1999). Biases in allocating obligations for climate protection: Implications from social judgement research in psychology. In F. Tóth (Ed.), *Fair weather: Equity concerns in climate change* (pp. 112–132). London: Earthscan.

Linneweber, V., Hartmuth, G., Deising, S., & Fritsche, I. (2001). Lokale Perspektiven globalen Wandels: Soziale Repräsentationen der Gefährdung Sylts angesichts möglicher Klimaänderungen [Local perspectives on global change: Social representations of the island Sylt facing potential climate change]. *Magdeburger Arbeiten zur Psychologie 3,(3).*

McDaniels, T., Axelrod, L. J., & Slovic, P. (1996). Perceived ecological risks of global change. A psychometric comparison of causes and consequences. *Global Environmental Change, 6*(2), 159–171.

Mesch, G. S., & Manor, O. (1998). Social ties, environmental perception, and local attachment. *Environment and Behavior, 30*(4), 504–519.

Moscovici, S. (1976). *La psychanalyse, son image et son public* [Psychoanalysis: Its image and its public] (2nd ed.). Paris: Presses Universitaires de France. (Original work published 1961).

Moscovici, S. (1984). The phenomenon of social representations. In R. M. Farr & S. Moscovici (Eds.), *Social representations* (pp. 3–69). Cambridge: Cambridge University Press.

Moscovici, S., & Duveen, G. (2001). *Social representations. Explorations in social psychology.* New York: New York University Press.

Norris Baker, C., & Scheidt, R. J. (1990). Place attachment among older residents of a 'ghost town': A transactional approach. In R. I. Selby, & K. H. Anthony (Eds.), *Coming of age.* (pp. 333–342). Edmond, Okla.: Environmental Design Research Association.

Pawlik, K. (1991). The psychology of global environmental change: Some basic data and an agenda for cooperative international research. *International Journal of Psychology, 26,* 547–563.

Perkins, D. D., & Long, D. A. (2002). Neighborhood sense of community and social capital: A multi-level analysis. In A. T. Fisher & C. C. Sonn (Eds.), *Psychological sense of community: Research, applications, and implications. The Plenum series in social/clinical psychology* (pp. 291–318). New York: Kluwer Academic/Plenum.

Proshansky, H. M. (1978). The city and self-identity. *Environment and Behavior, 10,* 147–169.

Proshansky, H. M., Fabian, A.-K., & Kaminoff, R. (1983). Place-identity: Physical world socialization of the self. *Journal of Environmental Psychology, 3,* 57–83.

Read, D., Bostrom, A., Morgan, M. G., Fischhoff, B., & Smuts, T. (1994). What do people know about global climate change? II. Survey studies of educated laypeople. *Risk Analysis, 14*(6), 971–982.

Reusswig, F. (1994). *Lebensstile und Ökologie. Gesellschaftliche Pluralisierung und alltagsökologische Entwicklung unter besonderer Berücksichtigung des Energiebereichs* [Lifestyles and ecology. Societal pluralization and ecological development in the energy sector]. Frankfurt: Verlag für Interkulturelle Kommunikation.

Ryan, R. L. (1998). Attachment to urban natural areas: Effects of environmental experience (Doctoral dissertation, Unversity of Michigan) *Dissertation Abstracts International, 58*(10A), 3764.

Selby, R. I., Anthony, K. H., Choi, J., & Orland, B. (Eds.). (1990). *Coming of age.* Edmond, Okla: Environmental Design Research Association.

Spencer, C., & Woolley, H. (2000). Children and the city: A summary of recent environmental psychology research. *Child: Care, Health and Development, 26*(3), 181–198.

Steel, G. D. (1995). The structure of the environmental relationship in polar regions (Doctoral dissertation, University of British Columbia). *Dissertation Abstracts International, 56*(4B), 2310.

Stern, P. C., Young, O. R., & Druckman, D. (1992). *Global environmental change: Understanding the human dimensions.* Washington, D.C.: National Academy Press.

Stokols, D., & Shumaker, S. A. (1981). People in places: A transactional view of settings. In J. H. Harvey (Ed.), *Cognition, social behavior and the environment* (pp. 441–488). Hillsdale, N.J.: Erlbaum.

Tajfel, H. (Ed.). (1978). *Differentiation between social groups: Studies in the social psychology of intergroup relations.* London: Academic Press.

Williams, D. R., Patterson, M. E., Roggenbuck, J. W., & Watson, A. E. (1992). Beyond the commodity metaphor: Examining emotional and symbolic attachment to place. *Leisure Sciences, 14*(1), 29–46.

III

Experiencing Nature as Members of Social Groups

12

Identity and Exclusion in Rangeland Conflict

Susan Opotow and Amara Brook

Jim and Patty ranch 10,000 acres at the foothills of the Rocky Mountains. Their land has been in Jim's family for generations, and he and Patty plan to pass it on to their children. They take pride in managing their land to maintain land and water quality and benefit wildlife, and they see their ranch as demonstrating that humans and nature can coexist to the benefit of both. Recently, however, nearby land has been split up and sold off in 40-acre homesites. It is disturbing that rangeland is disappearing, and it is frustrating that these new homeowners seem to know so little about managing their land. When Jim hears from a neighboring rancher that his ranch is within the habitat range for an endangered wildlife species, it elicits mixed feelings. On the one hand, Jim and Patty would be proud if their land benefited a rare species. On the other hand, they have heard about landowners being prevented from using their land as they wish because of endangered species listings. It strikes them as ironic that the very people who are moving to the Rockies for the unspoiled life-style, buying 40-acre lots and letting them become overrun with weeds, are the people pushing the endangered species listing.

In hundreds of communities in the eleven western states, ranchers, environmentalists, and federal land managers are wrestling with each other's differences in values, motives, needs, and cultures. Rangeland conflicts in the southwestern United States have sometimes been protracted and bitter. They primarily concern the management of public and private land and such related issues as water allocations, grazing leases, and species preservation. While it is a fictional composite, Patty and Jim's dilemma captures some of the ways that individuals experience conflict over rangeland management. It hints at the multiple incompatibilities, mutual obstructions, and social changes that suffuse environmental conflict. The vignette also emphasizes that environmental conflicts are rooted in particular histories, life-styles, hopes, and beliefs of individuals, families, and at larger levels of analysis, communities, groups, and societal institutions.

In this chapter we examine environmental identity within the context of an environmental conflict because, as we argue, environmental identities emerge, become salient in, and change in the course of environmental conflicts. We define environmental identity as having a personal and social component, and describe how differences in environmental identities can lead to moral exclusion, that is, seeing others and their well-being as morally irrelevant. We present data examining ranchers' environmental identity in a conflict over a threatened species listing in the southwestern United States, and we conclude with the implications of our findings for approaching environmental conflicts more constructively.

Facets of Environmental Identity

Because resource scarcity and environmental degradation (e.g., species endangerment and air pollution) are increasing (Annan, Flavin, & Starke, 2002; United Nations Development Programme et al., 2000), it is crucial to understand environmental identity—that is, how we see ourselves in relation to nature. A person's environmental identity consists of personal characteristics unique to the individual as well as group memberships shared with similar, like-minded others (Baumeister, 1998; Tajfel & Turner, 1986; Thomashow, 1995; Weigert, 1997). Personal identity includes such individual characteristics as intelligence, honesty, skills, and one's relationship with nature (e.g., being "environmentally concerned"; Baumeister, 1998). Mitchell Thomashow's (1995) definition of ecological identity as the individual's perceived interdependence with nature focuses on this personal aspect of environmental identity. Social identity is the portion of our self-concept that emerges from the perception that we share common characteristics, such as gender, ethnicity, political affiliations, or a similar orientation to nature (e.g., "environmentalist") with others (Tajfel & Turner, 1986; Turner, Hogg, Oakes, Reicher, & Wetherell, 1987). Andrew Weigert's (1997) definition of environmental identity as a socially shared understanding of the relationship between people and nature emphasizes the social aspect of environmental identity.

Although there is a tendency to perceive our personal and social identities as stable over time, they change as they are reinterpreted and imbued with meaning in specific social, physical, political, and cultural contexts (cf. Carbaugh, 1996; Cerulo, 1997; Hogg & Abrams, 1988; Markus & Kunda, 1986; Nagel, 1996).

Identity and Environmental Conflict

Environmental identity can contribute to the escalation of environmental conflict. A conflict exists "whenever incompatible activities occur. An action that is incompatible with another action prevents, obstructs, interferes, injures, or in some way makes the latter less likely or effective" (Deutsch, 1973, p. 10). Conflicts are constructive when they integrate multiple needs and foster creative problem solving and social change; conflicts are destructive when they seek to achieve benefits for oneself or one's group at the expense of others or, in escalating conflicts, when parties lose sight of their original goals and instead seek to inflict harm on others (Deutsch, 1973).

Environmental conflicts are essentially intergroup conflicts arising over such tangible resources as water or land (Klare, 2001) and such less tangible resources as political influence (Foa & Foa, 1974). In addition, environmental conflicts concern values, beliefs, and fairness (Clayton, 1994; Clayton & Opotow, 1994; Opotow, 1994). These conflicts can turn violent, such as in the White Mountains of Arizona, where environmentalists and federal land managers have opposed ranchers and loggers in conflicts over grazing leases, logging, recreation, and reintroduction of the Mexican wolf. Five wolves had been shot dead in the White Mountains by 1998 (Menning, 2000).

Intergroup conflict makes social identities, including environmental identities, more pronounced. Because an individual's self-esteem is partially rooted in self-enhancing between-group comparisons, people in competitive conflicts emphasize the positive characteristics of their own group (the ingroup) and the negative characteristics of groups with which they are in conflict (the outgroup), leading to outgroup distrust (Brewer, 1979; LeVine & Campbell, 1972; Luhtanen & Crocker, 1992). Given the opportunity to do so, people distribute more resources to their own group than other groups in order to enhance their own group's status and well-being (Brewer & Brown, 1998; Tajfel & Turner, 1986).

When social identity is paramount, people shift their focus from the level of the individual person to the level of the group and emphasize within-group similarities and between-group differences. As a result, people tend to stereotype, seeing their own group as "environmentalist" and another as "antienvironmentalist." This masks the fact that most people have an environmental identity that is more complicated than simply "for" or "against" something. Thus, environmental conflict makes competing social environmental identities more visible, fosters ingroup bias and stereotyping, and can exacerbate destructive conflict.

Moral Exclusion and Environmental Conflict

When environmental conflicts take a destructive course, individuals in one group may denigrate other groups and perceive them as outside their scope of justice, the psychological boundary within which moral norms, rights, and considerations of fairness apply (Deutsch, 1985; Opotow, 1990; Staub, 1990). When we exclude others from our scope of justice (i.e., morally exclude them), we see them as expendable, undeserving, exploitable, or irrelevant. Consequently, considerations of fairness do not apply to them; we are unwilling to allocate community resources to them and we are unwilling to make sacrifices that would foster their well-being (Opotow, 1987, 1993). When harm does befall them, it can seem inconsequential, inevitable, or deserved.

Moral exclusion is evident in environmental conflict in two ways. First, people can morally exclude other stakeholders. This can occur when people in an environmental conflict portray each other with such negative and morally charged labels as "ignorant," "selfish," and "zealot," to justify harms or losses experienced by those outside their scope of justice (also see Lerner, 1980). Second, people can morally exclude aspects of the natural world. This occurs when moral rules, values, and considerations of fairness do not apply, for example, to such disliked animals as bats or beetles (Opotow, 1987, 1993), or when people are unwilling to make sacrifices or allocate resources to protect wetlands, rivers, or the air from degradation (Opotow & Weiss, 2000).

The next section examines environmental identity and moral exclusion as it is evoked in rangeland conflict. It presents data on ranchers' perceptions of rangeland conflict and addresses three research questions:

1. How do ranchers characterize their own relationship to nature, others' relationship to nature, and their relationship to other stakeholders? That is, how is environmental identity evoked in this conflict? 2. How does environmental identity influence the course of this conflict? 3. Do ranchers morally exclude other people and nature, and do they think others morally exclude them? That is, how does moral exclusion emerge and influence this conflict?

Case Study: Rangeland Conflict

To examine environmental identity, we consider it in a specific case: the threatened species listing of the Preble's meadow jumping mouse (*Zapus hudsonius preblei*; Brook, 1999). The 1973 Endangered Species Act (ESA) has been a lightning rod for rancher–environmentalist conflicts in the West because conserving rare species requires careful management of both public and private land. On public lands, rangeland conflicts over the Endangered Species Act often center on grazing lease terms; environmentalists want to limit (through grazing fees) or eliminate cattle grazing on public lands to protect rare species and other aspects of environmental quality, while ranchers assert their right to use public land (Menning, 2000; Riseley, 1998). On private land, the ESA threatens landowners with potentially restrictive regulations. While key protagonists in rangeland conflicts are ranchers and environmentalists (including environmental groups), rangeland conflicts are multiparty conflicts that also concern governmental regulatory bodies such as the U.S. Fish and Wildlife Service and the Bureau of Land Management, indigenous populations, local people who do not depend on the rangeland for their livelihood, the larger public, and the nonhuman animate and inanimate natural world.

Against this backdrop, in 1998, the Preble's meadow jumping mouse ("Preble's") was listed as a threatened species under the ESA. The Preble's is not a well-liked, high-profile animal (such as the giant panda or the humpbacked whale) but instead is a small, prosaic rodent, a species that people tend to dislike (Kellert & Wilson, 1993) and actively eradicate as "vermin" and "a pest." The Preble's occupies riparian corridors on both private and public land from Colorado Springs, Colorado, in the south to Cheyenne, Wyoming, in the north. The species is threatened by accelerating agricultural, industrial, commercial, and residential development (Shenk, 1998; U.S. Fish and Wildlife Service, 1998).

In this case study, we focused on ranchers and their perspective on the Preble's listing since it affords an opportunity to understand the perspective of an important rangeland stakeholder whose view is not well understood in the environmental literature. We are aware that considerable diversity exists in the environmental orientation, expertise, and goals among rangeland stakeholders who are not ranchers, but because we focused on ranchers' perspective in this study, we labeled them as ranchers do, under the rubric, "nonranchers." This perception of homogeneity among outgroup members is, in fact, an aspect of intergroup conflict processes in action (Linville, Fischer, & Salovey, 1989).

Method

To explore ranchers' responses to the Preble's listing, open-ended interviews were conducted in June and July 1998 with thirteen owners of Preble's habitat, eleven of whom were ranchers (for details, see Brook, 1999). All eleven ranchers are men, reflecting the fact that males are more likely to own property. Interviews were open ended and dynamic in order to understand how landowners were framing the Preble's issue and what their concerns were. The questions included: When and how did you become aware that this species was listed as threatened under the ESA? What was your reaction? How did you think you personally would be affected by the listing? Have you changed your land management plans or practices in response to the listing? The interviews were analyzed using qualitative thematic content analysis (Boyatzis, 1998) to identify attitudes and concerns about the Preble's listing and behaviors in response to the listing.

These analyses were then used to construct a mail survey to test the distribution of attitudes and behaviors in the overall landowner population. (The survey contents and results are available from the authors.) The survey was sent to a random sample of 833 rancher and nonrancher owners of Preble's habitat stratified by county and clustered by section. The response rate was 46 percent ($N = 379$) after adjusting for invalid addresses. The survey respondents had an average age of 56; 73 percent were male, 33 percent had at least a four-year college education, 25 percent were employed in an agricultural occupation, and 48 percent had an income of more than $50,000 a year.

Results

In reporting both survey and interview findings, we focus on the ranchers' views of three aspects of this conflict: (1) the Preble's meadow jumping mouse; (2) wildlife, land, and nature management; and (3) fairness and societal structures. The first aspect samples environmental conflict in a specific, individual species context; the second aspect samples environmental conflict over habitat issues, a more general context; and the third aspect samples environmental conflict in a yet broader context, environmental regulation by the state and prevailing norms about fairness. Note that more specific contexts are nested within more general contexts.

Interview Findings
While the interviewers did not ask ranchers about their view of other groups, this information emerged spontaneously and forcefully, permitting us to examine how ranchers view nonranchers.

Preble's Meadow Jumping Mouse   The ranchers tended to view the Preble's as an insignificant species being championed by the very people responsible for its demise. One rancher remarked that there are "probably a lot of people wondering why they're even trying to save a mouse. We spend all our time trying to keep them out of our houses, and out of the grain bins, and out of this and out of this. Why all of a sudden are we worried about a mouse?" Another rancher speculated about causes of Preble's threatened status, saying "It seems to be the people who are living in the paved over areas who are (laugh) squawking about the thing. And the reason the mouse disappeared from where they are is because they built houses on it and are living there."

Some ranchers suspected that the threatened listing was based on questionable data. For example, one rancher stated "we're getting science, but nobody knows whether it's sound." Another rancher questioned the pending legislation: "It seems to me that whenever they've looked hard for it they find it so that bothers me a little bit. And maybe it's just wishful thinking but I'm suspicious that it's really threatened."

Managing Wildlife, Land, and Nature   One rancher described his relationship with the natural environment as both economic and aesthetic: "I love, I like wildlife as much as I like to harvest it. And I preserve it as much as I can." Another rancher said "ranchers don't have retirement plans, and so their retirement is the value of their land." Another added that "most agricultural people tend to be pretty good stewards of the land because if they aren't they're out of business."

The ranchers described themselves as responsible caretakers and stewards of natural habitat. One rancher who utilizes integrative management practices said, "These are practices that [I use to] try to control noxious weeds and things like that because I want the land to be healthy. . . . One of the jobs that any farmer has is to feed other people, and to do that you need healthy livestock and I'm convinced that to have healthy livestock or crops, you've got to take care of the land."

However, as another rancher pointed out, for ranchers, land stewardship has real costs:

Frankly, our deer and antelope harvest has been so poor that I won't let people hunt coyotes right now because I'd rather have the coyotes eat the antelope and deer if the hunters can't shoot them. . . . Four or five antelope eat as much as a cow so if you've got a hundred head out there it's just like having 20 cows you get no income from.

In contrast to their own expertise in land and wildlife management, ranchers saw nonranchers as inexperienced and irresponsible: "A lot of them are out there just because they don't like being in the city. And they don't know very much about the responsibilities of land stewardship." They also saw nonranchers as lacking solid knowledge:

I just think it's kind of crazy that they [environmental group] all of a sudden have expertise in ranching. Ten years ago they didn't think cows should be here. And all of a sudden now they claim to be the good, the example, of what proper ranching is.

Another rancher described outsiders' knowledge as insensitive to larger systems and cycles of change:

People love the beautiful green and if it turned black because of the fire you would think you had raped them or stole everything from them. They don't see the entire system. . . . If it catches on fire let it burn. I'm sorry folks but you can't have a golf course here. No you can't put your house here.

Another rancher described how poor land management practices of nonranchers affect them:

[New homeowners] don't manage at all. They've all got five acres, ten horses, and it's a mess. I mean it's bad. And, and other thing that's hurt is our weed control . . . the leafy spurge is eating this country alive. They think they're pretty yellow flowers but they don't know that's a very noxious weed that just keeps spreading and is going to eat this country. Which, I guess, if it's all going to be houses someday and they're happy to have the yellow flowers, then let them. It really drives you nuts when you try to control your weed problem and you can't do anything about the people right across the fence from you.

Ranchers saw nonranchers as invading and destroying rangelands: "They're taking some of the best country in the world and throwing houses on it." Another rancher described how nonranchers impose their rangeland aesthetic:

We'll get another bike path for a walk through the Preble's jumping mouse habitat restoration area. . . . I know I sound sarcastic, but . . . they'll build on the uplands so everyone can sit there on their deck and watch the Preble's jumping mouse with the red tail fox and say, 'Oh Martha! Wasn't that a wonderful one!'

Fairness and Societal Structures   The ranchers described themselves as making fair environmental decisions:

I certainly shoot coyotes if they're trying to kill my cattle it's sort of an economic decision. It's nothing against the coyotes. I always try to shoot to scare them off first. I give them a couple of chances to keep their distance and for the most part they do with just some warning shots. But I've had to kill them too if they don't give up and they come in and they start raising hell, there's no choice really at that point.

The ranchers saw nonranchers as disrespectful and procedurally unfair to them: "We get a little stuff forced on us along this line just because we are such a minority anymore. And we don't have much voice." Another rancher said:

Instead of working this thing out and saying, "Hey, let's see what we can do to get some habitat? Let's see what we need to do to protect this guy. Why don't we work on that?" All of a sudden, no, a big hammer comes out, as I call it, and we're going to put it on the Endangered Species or threatened species, and that's ridiculous. You know, I mean that is utterly ridiculous to handle that thing in that manner.

The ranchers described the regulatory burden as unfair: "Don't ask me to be the only person that pays the burden. . . . If the population in general wants to protect endangered species, which I think is great, then the population in general should be willing to pay for it." Another rancher explained, "There is such a strong sense of responsibility in the landowners, in ranchers. . . . They're caught in a tight spot. . . . The government comes along . . . and make[s] it [the mouse, a threatened species] a higher priority than my very livelihood."

Survey Findings

The survey findings, consistent with the interview data, emphasize that ranchers and nonranchers orient themselves differently to the Preble's meadow jumping mouse, to land management, and to fairness issues.

Preble's Meadow Jumping Mouse   The ranchers had a more negative view of Preble's than nonranchers, and this view corresponded to their behavior toward the mouse in some ways but not others. Ranchers were more biased against rodents in general and the Preble's mouse in particular than nonranchers ($t = 2.68$; $p < .01$). A greater proportion of ranchers (27 percent) than nonranchers (9 percent) managed their land to "minimize the chance of the Preble's living on it" ($\chi^2 = 13.05$; $p < .001$) and a smaller proportion of ranchers (25 percent) than nonranchers (53 percent) had allowed or, if asked, would allow a biological survey for the Preble's on their land ($\chi^2 = 15.40$; $p < .001$). Compared with nonranchers, ranchers were less supportive ($t = -3.46$, $p < .01$) of proposed regulations to protect the Preble's mouse and more supportive of exempting landowners from regulation ($t = 1.74$, $p < .10$). However, ranchers (28 percent) and nonranchers (21 percent) were equally likely ($\chi^2 = 1.14$; $p = .29$) to manage their land to improve Preble's habitat, and ranchers had more information about the Preble's than did nonranchers ($t = 5.96$, $p < .001$).

Managing Wildlife, Land, and Nature   Ranchers were as likely as nonranchers to think conserving wildlife is important, to enjoy having wildlife on their land ($t = 0.98$, $p = .33$), and to employ such conservation practices as constructing wetlands, modifying fences to allow wildlife migration, planting trees, raising native animals, and reducing fertilizer and herbicide use ($t = -0.35$, $p = .73$). Compared with nonranchers, ranchers were more concerned about their communities ($t = 6.03$, $p < .001$). The ranchers were also more likely to control weeds and manage their land using holistic conservation practices that link different parts of the range ecosystem ($t = 3.16$; $p < .01$). The ranchers were more likely than nonranchers to resent being told how to manage their land ($t = 2.66$; $p < .01$) and to agree with the statement, "If the Preble's meadow jumping mouse exists on my land then I must already be doing the right things to protect it" ($t = 4.31$; $p < .001$), indicating their confidence in their land and wildlife management expertise.

Fairness and Societal Structures   Compared with nonranchers, ranchers were more likely to feel negatively toward the federal government ($t = 4.44$; $p < .001$) and to prefer landowner and local control of land over federal control ($t = 4.57$; $p < .001$). Consistent with this, 92 percent of the ranchers agreed that "environmentalists don't understand the pressures that landowners face" and 88 percent disagreed that "landowners should bear financial responsibility for achieving public conservation goals on their land." Compared with nonranchers, ranchers were more worried that the ESA would not be enforced equitably ($t = 2.47$; $p < .001$).

Identity and Moral Exclusion in Rangeland Conflict

What do these data tell us about the meaning of environmental identity as it is evoked in environmental conflict? The ranchers' views indicated that differences in environmental identity can fuel intergroup conflict and lead to moral exclusion.

Environmental Identity
Ranchers perceive their orientation toward the natural world as different from that of nonranchers (table 12.1). They believe that compared with nonranchers, they have different views of the natural world, different goals for land management, and different responses to governmental land regulations.

Ranchers' Environmental Identity   Ranchers enjoy having wildlife on their land. Because they are directly dependent on nature's whims, they see nature as both good and bad: "Ranchers see themselves as being *in* nature rather than outside looking in, whereas more often urban environmentalists seem often only to feel they are *inside* of nature when they

Table 12.1
From the ranchers' perspective: Contrasts between ranchers and others' environmental identity

Ranchers see themselves	Ranchers see others
Coexisting with and depending on nature	Living apart from nature
Encouraging and supporting wildlife	Destroying natural habitat through overdevelopment
As knowledgeable and careful land stewards	Lacking knowledge about systems and holistic land-management practices
Utilizing time-tested, integrative management practices	Utilizing species protection as an excuse for imposing coercive regulations on ranchers
As recipients of economic and regulatory burdens for species and rangeland protection	As having a coercive rather than a collaborative approach to rangeland management

are in a recreational, wilderness setting" (Riseley, 1998, p. 86). Because ranchers see themselves as part of nature, coexisting with and depending on nature as well as encouraging and supporting wildlife, they believe that humans and nature can coexist within the same landscape (Peterson & Horton, 1995; Riseley, 1998). A Colorado rancher summarized this idea, "I think that wildlife habitat protection can be done on most agricultural operations." Because of their direct interaction with nature, they pride themselves on being knowledgeable land stewards who utilize time-tested, integrative management practices. Instead of being appreciated for their knowledge, skills, and effort, however, they see themselves as punished with undue economic and regulatory burdens for species and rangeland protection. They see regulatory rules and processes as disrespectful and dictatorial because they establish the government (and sometimes environmental groups) as experts and demand that ranchers conform to rules, undermining their environmental expertise and identity.

Ranchers' Views of Other Stakeholders' Environmental Identity   The ranchers' view of nonranchers' environmental identity is largely consistent with the identity that nonranchers themselves espouse. Self-identified environmentalists, for example, have great affection for nature but see humans as outside of it (Riseley, 1998), believe that nature should be protected as wild and pristine, and urge that the rights of nature should be respected by all people (Nash, 1989). From this perspective, allowing cattle to forage with wildlife on rangeland is environmental degradation (Larmer, 1996).

Ranchers become very frustrated with those who demand sacrifice from resource users to foster land and wildlife conservation but at the same time deny that they, too, depend on nature to meet their needs. The environmentalists are perceived as hypocritical, as described by a woman rancher from South Dakota who states that they "continue to wear leather shoes, lust after leather upholstery, and relish a good hamburger . . . but remain unwilling or unable to slaughter and skin the animals that can satisfy them" (Riseley, 1998, p. 86, quoting Hasselstrom, 1992). Another aspect of environmentalists' hypocrisy, as ranchers see it, is illustrated in figure 12.1 (2001): A self-proclaimed "nature lover" asserts his self-serving and idiosyncratic view of nature selfishly, forcefully, and callously. McKie's cartoon captures ranchers' view of nonranchers as concerned about nature only on their own, ill-informed terms while ignoring the impact that their behavior has on nature.

Ranchers' environmental identity also has an oppositional component (Ogbu, 1991) consisting of disapproval, dislike, and distrust of nonranchers and their environmental agenda. Although ranchers' self-identify themselves as deeply proenvironmental because they conserve nature of their own volition, they are leery of supporting any kind of regulation, even when it would foster wildlife or land protection. This reluctance to support proenvironmental initiatives leads others to characterize them as antienvironmental. Ranchers are unhappy about this characterization. A Colorado rancher described how he integrates conservation issues and his environmental identity:

We really try to coexist with the wildlife. We really enjoy having it around. We're very careful about how we graze along the creek and all that, looking at the erosion and stuff like that. So on the one hand you know, I think we're pretty good stewards of the land and it's very possible that that mouse is there and living quite happily. I guess the fear that anybody, a rancher has is if, you may be doing just the right things and the mouse may be happy but suddenly there might be some federal regulations that will tell you have to do more or that you can't graze on this land or you can go near the creek within a certain distance,

Figure 12.1
Ranchers' view: environmentalists ignore the impacts of their own behavior.
(Drawing by C. H. McKie.)

or something . . . you really kind of hope that the animal is not on your land,
which is kind of contrary to my view of how I like to have open land and be a
steward of the land, because I like to have all wildlife and encourage it. And yet
if suddenly there's some wildlife there that restricts your use of the land in a
major way, you kind of hope that they aren't there. And it's too bad that a
rancher is put in that position, of not wanting wildlife, when in fact most are
quite supportive of wildlife.

Social Identity Processes

While conflict between ranchers and environmentalists is partially attrib-
utable to real differences in environmental identities, it is also exacer-
bated by such social identity processes such as stereotyping, distrust,
suspicion, and discrimination. Ranchers' social identity is increasingly
threatened as environmentalists gain political influence in the southwestern United States while ranchers and farmers are losing their influence (Menning, 2000; Rothman, 2000).

In the context of environmental conflict, stereotypes color interpretations of the others' behavior and can intensify fear. For example, due to concern that overgrazing may harm rare species habitat and damage streams and vegetation, environmentalists sometimes argue that cattle do not belong on public or private land. Ranchers not only see this as threatening their livelihoods but also as devaluing them as individuals and a group because ranchers' well-being is placed below that of nature—even a mouse. As one rancher pointedly stated, "The government comes along ... and [makes the mouse] a higher priority than my very livelihood." Ranchers see pending government action as opposing them and their interests, and they see groups with more political power and resources than they have—the government, environmentalists, and others— arrayed against them. As McKie (1999) illustrates in figure 12.2, ranchers see these opponents colluding against them, utilizing sophisticated technological resources and the Preble's mouse as fodder in a battle to halt ranching.

Figure 12.2
Ranchers' view: the Preble's meadow jumping mouse as a weapon in the battle against ranching. (Drawing by C. H. McKie.)

Commenting on their feelings regarding the listing of the Preble's as threatened, a Colorado rancher stated:

What we think we know is what the government and Fish and Wildlife Service have done toward halting different activities where there's a habitat threatened. And so, that is they're capable because they've done some things. . . . So we're believing the worst. . . . I think Fish and Wildlife folks have tried to point out that agriculture itself is not as big an intrusion on the mouse's habitat as some other things. . . . But the information, I think, is not complete and landowners are as bad as anyone else about jumping to conclusions, because you feel the worst, fear the worst.

This fear and threat can lead to distrust, suspicion, the feeling of being discriminated against, and ultimately, exclusion of the opposing group from one's scope of justice.

Moral Exclusion

In rangeland conflicts, social and environmental identities and beliefs about conservation practices can clash and give rise to the moral exclusion of other rangeland stakeholders as well as the natural world. Ranchers' and nonranchers' differing social and environmental identities can lead them to demonize each other and not see other stakeholders as within their moral universe. Many nonranchers are drawn into rangeland conflicts because they love nature (Riseley, 1998), but ranchers trivialize them as "squawking" people from "paved-over areas" who destroy natural habitat "because they built houses on it" or malign them as power-hungry, ignorant, green extremists. Although these data capture only ranchers' dismissive and exclusionary views, some nonranchers also see ranchers in similar derogatory, terms: exploitative, antienvironmentalists, and private-property rights zealots, and as "welfare parasites who transform the soil and grass into dust and weeds at taxpayer expense" (Riseley, 1998, p. 28). When ranchers and other resources users who depend on nature for their livelihood and live in it (e.g., loggers) are viewed as having no right to make a living from the land, it is only a short step to seeing their needs, concerns, and well-being as irrelevant in planning conservation initiatives. Even when ranchers are drawn into local community efforts, it may happen because environmental groups feel that this is the only practical way to make conservation happen, not because ranchers are seen as legitimate stakeholders who ought to be included.

In addition to morally excluding each other, stakeholders in rangeland conflicts also can morally exclude the natural world. Differences in orientations toward nature and beliefs about appropriate land and wildlife management practices yield different inclusionary beliefs and priorities for ranchers and nonranchers. While ranchers do not generally include the Preble's in their scope of justice ("We spend all our time trying to keep them out of our houses, and out of the drain bins, and out of this and out of this. Why all of a sudden are we worried about a mouse?") they do see themselves as including all of nature within their scope of justice and make considerable sacrifices to do so. Ranchers view nonranchers as including the Preble's and all of nature within their scope of justice, but doing so hypocritically, because of their unwillingness to make real sacrifices in personal life choices and instead heaping responsibility on ranchers to protect the Preble's and nature.

As table 12.2 shows, from a rancher's perspective, ranchers and nonranchers agree in their exclusion of each other, but their inclusion of the natural world (i.e., Preble's and nature) is quite distinctive for each party. As a result, each side sees ways that they include the Preble's mouse and nature and ways that the other side excludes them. Consequently, each

Table 12.2
Moral exclusion (from the ranchers' perspective)

Attitude	Ranchers see themselves as including			Ranchers see nonranchers as including		
	Preble's	Nature	Nonranchers	Preble's	Nature	Ranchers
Perceiving fairness as applicable to:	No	Yes	No	Yes	Yes	No
Willingness to allocate resources to:	Yes/No	Yes	No	Yes	Yes	No
Willingness to make sacrifices that foster the well-being of:	Yes/No	Yes	No	No	No	No

side claims the high moral ground, dubbing itself "environmentalist" and the other side "antienvironmentalists" and "hypocrites." Partial, compartmentalized moral inclusion and exclusion along with oppositional environmental identities justifies these labels and makes them appear apt.

Implications for Practice: Resolving Rangeland Conflict Constructively

As our data have indicated, ranchers describe themselves as responsible stewards attuned to and part of nature, while they describe nonranchers as ignorant, irresponsible, insincere, and separate from nature (see table 12.1). Through selective attention and denial ranchers and environmentalists construct simplified, mythic environmental identities for themselves and others (cf., Opotow & Weiss, 2000; Riseley, 1998). These stereotypical environmental identities are defensive on both sides. They deny the ranchers' culpability in environmental degradation while at the same time they deny the positive aspects of nonranchers' environmental behavior. These processes occur not only for ranchers but also among other parties in environmental disputes over the management of natural resources throughout the world. What do our data on identity suggest about more constructive approaches to these conflicts?

Environmental conflicts concerning the use and control of natural resources are often depicted as rational, real struggles to achieve concrete objectives. Such "rational actor" explanations argue that each stakeholder seeks to maximize its self-interests and goals (Baumol & Blinder, 1988; Sherif, 1966). "Rational actor" explanations assume that self-identified environmentalists are motivated to conserve nature, aesthetic beauty, and recreational activities, while ranchers are motivated to protect their financial interests. Such analyses suggest that if the groups could agree on a solution that satisfied these real interests of both parties, the conflict could be resolved (Fisher & Ury, 1981). However, this approach does not always work. Why not? Analyses assuming that behavior in resource conflicts is motivated by the desire of each side to satisfy its substantive interests miss some important dynamics of the situation. These dynamics are captured by understanding the social aspects of environmental identity.

Social identity theory proposes that an additional reason for intergroup conflict, beyond competition for real resources, is social competition

(Turner, 1975). A "rational actor" model predicts that groups should welcome a "win-win" solution that increases the well-being of all parties. However, when faced with decisions about resource allocations, groups prefer seemingly irrational outcomes that maximize differences between themselves and other groups, rather than an outcome that maximizes the benefit to their own group (Tajfel, 1982; Billig & Tajfel, 1973). This occurs because groups assess their collective self-worth by comparing themselves with other groups (Hogg & Abrams, 1988; Luhtanen & Crocker, 1992). Thus, social competition helps explain why ranchers' and environmentalists' conflicts over rangeland, including land management, species protection, and water allocation issues, can be so difficult to resolve. As our data indicate, environmental conflict elicits and fosters distinctive and oppositional environmental identities. These, in turn, foster the moral exclusion of opposing groups. As a result, part of the "real interests" of parties in conflict is to protect their dignity in the face of perceived disrespect, derogation, and moral exclusion as much as to maintain control of a resource they see as crucial to their well-being.

Two approaches address identity-based impediments to conflict resolution. The first suggests that creating an overarching, larger, inclusive ingroup can promote conflict resolution by decreasing the between-group biases that feed opposing environmental identities in environmental conflict (cf., Allport, 1954; Gaertner, Dovidio, Anastasio, Bachman, & Rust, 1993). Creation of such overarching identities, such as "preservers of open space," has led to effective collaboration in locally led rangeland and watershed planning efforts in the Southwest. (See, for example, the Quivera Coalition [http://www.quiviracoalition.org] and the Diablo trust [http://www.for.nau.edu//diablo_trust].)

Creating an overarching identity, however, is not enough. A second approach advises that, while creating an overarching identity can reduce conflict, it is also important to preserve subgroup identities for several reasons. First, preserving subgroup identities can increase the generalizability of constructive attitudes and conflict reduction processes beyond the immediate context (Gaertner et al., 1993). For example, in a collaborative process, ranchers may build trusting relationships with particular environmentalists, and vice versa. If these people continue to see each other as members of the groups "ranchers" and "environmentalists," they are more likely to extend goodwill to other ranchers and environmentalists than if they see the other person only as an individual, unlike others in their group. As a consequence, if subgroup identities are preserved, ranchers might reduce their prejudice toward all "environmentalists" and not just those in their local area with whom they now interact; similarly, nonranchers might reduce their prejudice toward all ranchers or resource users and not just those in their local area with whom they now interact.

A second reason to preserve subgroup identities is that failure to do so may make participants feel that their original identities are threatened or lost (Hewstone, 1996; Hewstone & Brown, 1986; Hornsey & Hogg, 2000a). Consequently, stakeholders might then behave in a more discriminatory fashion than they did originally (Hornsey & Hogg, 2000b). Preserving and valuing subgroup identities may be especially important in rancher–environmentalist conflicts because each group already sees the other as threatening its identity (Riseley, 1998).

A third reason to preserve subgroup identities is that people have an ideal level of distinctiveness from and similarity to other people. They want to be good members of their ingroups and do not want to lose their subgroup identities in superordinate groups (Brewer, 1991). This suggests the importance of appreciating ranchers' expertise, gleaned over years of work, as well as sacrifices they have made toward conservation of land, plants, or animals. Mary Burton Riseley (personal communication, 2001) proposes that "appreciative inquiry" rather than oppressive dictates or punitive measures would be a more constructive approach by officials, agents, and others who administer public grazing lands (e.g., Bureau of Land Management and state land offices). This would not only produce a more constructive approach to conflicts, but could also foster moral inclusion. Moral inclusion is the recognition, not the appropriation or obliteration, of subgroup identities. Inclusion in action means operationalizing constructive obligations to subgroups by applying moral norms, rights, and considerations of fairness to them; allocating resources to them; and even making sacrifices that would shift the prevailing status quo in order to do so (Opotow, 2001).

In sum, just as rangelands are complex ecologically, conflict dynamics among human stakeholders are also complex. And just as interdependence characterizes natural habitats, the recognition of their interdependence is also crucial for the many human stakeholders that live on, love, benefit from, and gain their sense of identity from rangelands.

References

Allport, G. (1954). *The nature of prejudice*. Reading, Mass.: Addison Wesley.

Annan, K. A., Flavin, C., & Starke, L. (2002). *State of the world, 2002: A Worldwatch Institute report on progress toward a sustainable society*. New York: Norton.

Baumeister, R. F. (1998). The self. In D. Gilbert, S. Fiske, & G. Lindzey (Eds.), *The handbook of social psychology* (Vol. 1, pp. 680–740). Boston: McGraw-Hill.

Baumol, W. J., & Blinder, A. S. (1988). *Economics: Principles and policy*. San Diego: Harcourt Brace Jovanovich.

Billig, M., & Tajfel, H. (1973). Social categorization and similarity in intergroup behaviour. *European Journal of Social Psychology, 3*(1), 27–52.

Boyatzis, R. E. (1998). *Transforming qualitative information: Thematic analysis and code development*. Thousand Oaks, Calif.: Sage.

Brewer, M. B. (1979). The role of ethnocentrism in intergroup conflict. In W. G. Austin & S. Worchel (Eds.), *The social psychology of intergroup relations* (pp. 71–84). Monterey, Calif.: Brooks-Cole.

Brewer, M. B. (1991). The social self: On being the same and different at the same time. *Personality and Social Psychology Bulletin, 17*, 475–482.

Brewer, M. B., & Brown, R. J. (1998). Intergroup relations. In D. Gilbert, S. Fiske, & G. Lindzey (Eds.), *The handbook of social psychology* (Vol. 2, pp. 554–594). Boston: McGraw-Hill.

Brook, A. (1999). Landowner responses to an Endangered Species Act listing: Some recommendations for encouraging private land conservation. Master's thesis, University of Michigan, Ann Arbor.

Carbaugh, D. (1996). *Situating selves: The communication of social identities in American scenes*. Albany, NY: State University of New York Press.

Cerulo, K. A. (1997). Identity construction: New issues, new directions. *Annual Review of Sociology, 23*, 385–409.

Clayton, S. (1994). Appeals to justice in the environmental debate. *Journal of Social Issues, 50*(3), 1913–1927.

Clayton, S., & Opotow, S. (Eds.). (1994). Green justice: Conceptions of fairness and the natural world. *Journal of Social Issues* [Special issue] *50*(2).

Deutsch, M. (1973). *The resolution of conflict; constructive and destructive processes*. New Haven, Conn.: Yale University Press.

Deutsch, M. (1985). *Distributive justice: A social-psychological perspective*. New Haven, Conn.: Yale University Press.

Fisher, R., & Ury, W. (1981). *Getting to yes: Negotiating agreement without giving in*. Boston: Houghton Mifflin.

Foa, U. G., & Foa, E. B. (1974). *Societal structures of the mind*. Springfield, Ill.: Charles C Thomas.

Gaertner, S. L., Dovidio, J. F., Anastasio, P. A., Bachman, B. A., & Rust, M. C. (1993). The common ingroup identity model: Recategorization and the reduction of intergroup bias. *European Review of Social Psychology, 4*, 1–26.

Hasselstrom, L. M. (1992). Letter to the editor. *Garbage, 4*(3), 18–19.

Hewstone, M. (1996). Contact and categorization: Social psychological interventions to change intergroup relations. In C. N. Macrae, C. Stangor, & M. Hewstone (Eds.), *Stereotypes and stereotyping* (pp. 323–368). New York: Guilford.

Hewstone, M., & Brown, R. (1986). Contact is not enough: An intergroup perspective. In M. Hewstone & R. Brown (Eds.), *Contact and conflict in intergroup encounters* (pp. 1–44). Oxford: Blackwell.

Hogg, M. A., & Abrams, D. (1988). *Social identifications: A social psychology of intergroup relations and group processes*. Florence, Ky.: Taylor & Francis/Routledge.

Hornsey, M. J., & Hogg, M. A. (2000a). Assimilation and diversity: An integrative model of subgroup relations. *Personality and Social Psychology Review, 4*(2), 143–156.

Hornsey, M. J., & Hogg, M. A. (2000b). Subgroup relations: A comparison of mutual intergroup differentiation and common ingroup identity models of prejudice reduction. *Personality and Social Psychology Bulletin, 26*, 242–256.

Klare, M. T. (2001). *Resource wars: The new landscape of global conflict*. New York: Metropolitan Books.

Kellert, S. R., & Wilson, E. O. (1993). *The biophilia hypothesis*. Washington, D.C.: Island Press.

Larmer, P. (1996). Judge sends a message to cows. *High Country News*, Oct. 28, p. 28.

Lerner, M. J. (1980). *The belief in a just world: A fundamental delusion*. New York: Plenum.

LeVine, R. A., & Campbell, D. T. (1972). *Ethnocentrism: Theories of conflict, ethnic attitude, and group behavior*. New York: Wiley.

Linville, P. W., Fischer, G. W., & Salovey, P. (1989). Perceived distributions of the characteristics of in-group and out-group members: Empirical evidence and a computer simulation. *Journal of Personality and Social Psychology, 57*(2), 165–188.

Luhtanen, R., & Crocker, J. (1992). A collective self-esteem scale: Self-evaluation of one's social identity. *Personality and Social Psychology Bulletin, 18*(3), 302–318.

Markus, H., & Kunda, Z. (1986). Stability and malleability of the self-concept. *Journal of Personality and Social Psychology, 51*(4), 858–866.

McKie, C. H. (1999). Have mouse will travel. *Elbert County News*, Jan. 21, p. 6A.

McKie, C. H. (2001). Cherry creek trail. *Douglas County News-Press*, Feb. 14, p. 6A.

Menning, N. L. (2000). Residents and visitors: Defining people-forest relationships in the new west. Paper presented at the 8th International Symposium on Society and Resource Management.

Nagel, S. A. K. (1996). Conflict and harmony: Linkages between the quality of parents' marital interactions and the quality of their children's sibling interactions. *Dissertation Abstracts International: Section B: The Sciences & Engineering, 56*(12–B).

Nash, R. (1989). *The rights of nature: A history of environmental ethics.* Madison: University of Wisconsin Press.

Ogbu, J. U. (1991). Minority coping responses and school experience. *Journal of Psychohistory, 18*(4), 433–456.

Opotow, S. (1987). Limits of fairness: An experimental examination of antecedents of the scope of justice. *Dissertation Abstracts International, 48*(08B), 2500.

Opotow, S. (1990). Moral exclusion and injustice: An introduction. *Journal of Social Issues, 46*(1), 1–20.

Opotow, S. (1993). Animals and the scope of justice. *Journal of Social Issues, 49*(1), 71–85.

Opotow, S. (1994). Predicting protection: Scope of justice and the natural world. *Journal of Social Issues, 50*(3), 1949–1963.

Opotow, S. (2001). Reconciliation in times of impunity: Challenges for social justice. *Social Justice Research, 12*(2), 149–170.

Opotow, S., & Weiss, L. (2000). Denial and the process of moral exclusion in environmental conflict. *Journal of Social Issues, 56*(3), 475–490.

Peterson, T. R., & Horton, C. C. (1995). Rooted in the soil: How understanding the perspectives of landowners can enhance the management of environmental disputes. *Quarterly Journal of Speech, 81*(2), 139–166.

Riseley, M. B. (1998). Ranchers and environmentalists collaborate: Learnings from a corner of the Southwest. Master's thesis, University of Massachusetts, Boston.

Rothman, H. (2000). Do we really need the rural west? *High Country News*, April 24, p. 32.

Shenk, T. (1998). *Conservation assessment and preliminary conservation strategy for Preble's meadow jumping mouse (Zapus hudsonius preblei).* Fort Collins, Cal.: Colorado Division of Wildlife.

Sherif, M. (1966). *The psychology of social norms.* New York: Harper Torchbooks.

Staub, E. (1990). Moral exclusion, personal goal theory, and extreme destructiveness. *Journal of Social Issues, 46*(1), 47–64.

Tajfel, H. (1982). Social psychology of intergroup relations. *Annual Review of Psychology, 33*, 1–39.

Tajfel, H., & Turner, J. C. (1986). The social identity theory of intergroup behavior. In S. Worchel & W. Austin (Eds.), *Psychology of intergroup relations* (pp. 7–24). Chicago: Nelson Hall.

Thomashow, M. (1995). *Ecological identity: Becoming a reflective environmentalist*. Cambridge, Mass.: MIT Press.

Turner, J. C. (1975). Social comparison and social identity: Some prospects for intergroup behaviour. *European Journal of Social Psychology, 5*(1), 5–34.

Turner, J. C., Hogg, M. A., Oakes, P. J., Reicher, S. D., & Wetherell, M. S. (1987). *Rediscovering the social group: A self-categorization theory*. Oxford: Blackwell.

United Nations Development Programme, the United Nations Environmental Programme, the World Bank, and the World Resources Institute. (2000). *World resources 2000–2001: People and ecosystems: The fraying web of life*. Washington, D.C.: World Resources Institute.

U.S. Fish and Wildlife Service. (1998). Endangered and threatened wildlife and plants: Final rule to list the Preble's meadow jumping mouse as a threatened species. *Federal Register, 63*(92), 26517–26530.

Weigert, A. J. (1997). *Self, interaction, and natural environment*. Albany: State University of New York Press.

13

Group Identity and Stakeholder Conflict in Water Resource Management

Charles D. Samuelson, Tarla Rai Peterson, and Linda L. Putnam

There is an amorphous body of knowledge that makes little or no difference in the daily conduct of our lives. By contrast, some knowledge seems to demand action. For example, over the past 50 years, ecologists, hydrologists, and others have conducted extensive research on water quality in urban centers throughout the United States. These scientists have agreed that some harmful substances, such as mercury, are concentrated in the tissues of various plants and animals. The information that mercury levels in an urban watershed exceed the amounts known to have caused symptoms of methyl mercury poisoning (i.e., dysfunction in limbs, speech, vision, and mental capacity) in another population requires a behavioral response. Knowledge that relates problems to persons, interests, and actions often implies an imperative for choice and action.

The central problem addressed in this chapter is: When does knowledge about environmental problems result in action and when does it not? We propose that group identity may be an important mediator between knowledge and action. Figure 13.1 shows our theoretical framework for analyzing the links between knowledge and action. In this chapter we use this conceptual model to organize our presentation of two case studies on water resource management.

Overview of Theoretical Framework

The chapter is organized around two focal relationships. First, the relationship between knowledge and group identity is assumed to be indirect, mediated through three antecedent processes (solid arrows in figure 13.1) that contribute to the development of group identity: (1) consensus attribution, (2) group interaction, and (3) generative process.

Second, the relationship between group identity and action is also predicted to be indirect, with mediation via identity frames and normative influence operating independently, as illustrated by the two solid arrows between these mediators and action.

In addition to these focal relationships, figure 13.1 contains recursive feedback loops (denoted by dashed arrows) between action and group identity, between action and normative influence, and between identity frames and group identity. Actions taken by the group are expected to reinforce the sense of group identity, particularly those actions that support the group's core beliefs and values. Actions that conform to the group's behavioral norms should increase the normative influence of group identity by clarifying those behaviors that are approved or disapproved for group members. Finally, identity frames are expected to feed back on group identity in a reciprocal process because members' interpretative frames will guide the search for and processing of information to further reinforce group identity.

The remainder of this section elaborates on the relationships depicted in figure 13.1. We turn first to defining what we mean by the term *group identity*.

Group identity refers to how an individual answers the question "Who am I?" as it is linked to the group, organization, or community to which this person belongs (Hoare, 1994; Roland, 1994). Similarly, Tajfel and Turner (1985) view group identity as a self-image created through

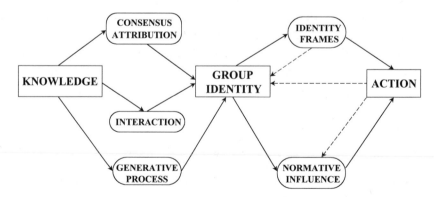

Figure 13.1
Conceptual framework for hypothesized relationships among knowledge, group identity, and action.

membership in a social category. Identity is iterative, shaping and being shaped by group membership, social values, and cultural experiences. Although individuals conceptualize their identities in a variety of ways, people often describe themselves through the use of a social or institutional label, such as an environmentalist, logger, or agency official (Hogg, Terry, & White, 1995).

More specifically, the construct of identity is used in two distinct yet related senses in this chapter. First, the term *identity frames*, following Gray's (1997) typology, is introduced to capture the interpretive, sense-making function of group identity. Second, we use the term *group identity* in our analysis of a local watershed restoration council, to refer to the participants' motivation and competence to participate as a cohesive group in environmental management decisions. In this latter context, group identity is conceptualized through a social constructionist lens; we argue that this form of identity evolves gradually over time through communicative exchanges that occur among group members. The next section develops the theoretical basis of this argument.

Group Identity as Consensus Attribution

The concept of consensus, broadened by Scheff (1967) to include an awareness that agreements are held, is central to group identity. Group identity rests upon a form of consensus that is attributed rather than shared. This means that the assumption of agreement may be wrong. It is an assumed understanding of agreement, rather than its fact, that creates sufficient identification for the group to act.

To illustrate this characteristic of group identity, recall the mercury poisoning example introduced earlier. The outcome of this research on mercury contamination may be understood as specialized or technical knowledge. This knowledge was based upon a real consensus among scientists as to appropriate research methods and measurement techniques. This consensus does not protect such technical knowledge from sources of error. However, even the determination and correction of error requires an underlying methodological consensus.

By contrast, the consensus of any segment of the lay public on the significance, seriousness, or harm posed by mercury contamination must be attributed to the group in order to initiate management activities. Technical knowledge reflects the outcome of an actual consensus on specialized research procedures. Group identity, on the other hand, may accomplish the same thing by attributing a consensus concerning the generalized interests of persons within the group (e.g., desire to protect the quality of life in a community from the harmful health effects of mercury contamination).

If group identity is to be meaningful, at least two factors must be present: (1) the group must have the opportunity to refrain from acting and (2) it must have a reasonable expectation that its chosen action can change the situation.

Group Identity as Interaction

Participants assume that other members of their group share their identity. For information to function communicatively, it must depend upon basic assumptions of public consensus on certain problems, interests, and actions. If a particular advocate notes, for example, that mercury concentrations in fish from an inner-city waterway are dangerously high, this person claims a pragmatic faith in the mutuality of social interests (i.e., the value of living in a healthy environment that is free from mercury contamination).

This faith cannot be empirically verified, nor will it always be well founded. But if communication is to operate effectively, some "beginning" must exist, whether in empirical fact or assumption. The assumption is grounded in the possibility that those who play the collective role of public may become conscious that the suffering of others is relevant to their own interests. An advocate does not need to assume that the target audience has a technical comprehension of water quality, but she or he does assume that there is an appreciation of human potentials and skills within society and the relevance of these to the purposes of a community. In proposing a solution to inner-city water pollution, the advocate assumes, at a minimum, that there is some conception of pollution in relation to the social interest.

Group identity exists in a state of potential until, through the reasoned action of the group itself, that potential state of identity is realized. Just as the specialized consensus on methods of investigation has been validated through repeated scientific operations, so is group identity affirmed through recurrent interaction.

Group Identity as a Generative Process

Rather than being fixed, group identity is transitional and generative. As individual problems are encountered and managed through the frustrating incrementalism of human decision making, new problems emerge and with these, new knowledge may be attributed, based reasonably upon past collective judgments. Group identity thus establishes social precedents for future attributions of consensus in situations that have yet to be encountered.

This generative characteristic of group identity can be illustrated in the development of the traditional issues of controversy. The holders of two or more opposing positions may, at various times, reduce their differences to a question of fact, definition, quality, or procedure (Prelli, 1989). That question, when settled, may determine the direction and eventually the outcome of the conflict. Yet if this process of controversy is to operate effectively, we must assume there is consensus on a prior issue in order to move properly to the next. For instance, we may not argue over the distinguishing characteristics of the pollution crisis in any urban center, or the seriousness of its effects upon relevant human interests, unless we attribute to our audience a prior consensus on the presence of that crisis. Thus, the components of group identity (i.e., assertions of fact, definitions of character, rules of quality, or precedents of procedure) aid each communicative exchange in its completion. When a controversy reaches a point of temporary resolution, a more fully actualized consensus is achieved that functions as a social precedent for future behavior.

These three antecedents to group identity formation contribute to an understanding of its normative implications for action. Once group identity has emerged through attributed consensus and group interaction in a generative process, strong pressures are brought to bear on members to conform to specific group norms. Thus, group identity generates an imperative to act in ways consistent with the attributed consensus on issues important to the group's values and goals.

Figure 13.1 also suggests that group identity can influence action through the construct of identity frames.

Identity Frames and Action in Environmental Disputes

Framing has become an important construct in understanding conflict management in general (Putnam & Holmer, 1992) and environmental disputes in particular (Gray, 1997; Taylor, 2000). Frames refer to the worldviews, interpretations of experience, and perspectives that parties bring to situations. They help parties determine what is figure and what is ground. They provide participants with accounts for what is included or excluded, for ways of acting and reacting, and for interpreting the actions of other parties in the dispute. This definition of framing is rooted in social dynamics and interpretations of events, a feature that distinguishes it from treating frames as cognitive models of judgment (cf. Bazerman & Neale, 1992; Tversky & Kahneman, 1974).

Recent literature on environmental conflict resolution has emphasized the importance of identity frames. Disputants often adopt multiple and seemingly contradictory identities; for example, "stewards" of land as well as "victims" of land use. These characteristics become part of an individual's frame and influence what they should believe as well as how they should behave (Ashforth & Mael, 1989; Taylor & Moghaddam, 1994). Gray (1997), building on Douglas and Wildavsky's (1982) work on risk perception, argues that the frames used by stakeholders to understand and interpret environmental conflicts are often based on personal and/or group identities. Gray (1997) proposes that environmental disputes may be driven as much by conflicts over identities among stakeholders as they are by specific issues and interests that divide the parties.

Such conflicts are often generated when environmental disputants experience threats to their identities (Carbaugh, 1996). This is true for all groups, including environmentalists, loggers, and government officials. For example, the identities and livelihoods of some loggers are jeopardized by environmental restrictions that prevent logging on public lands. Similarly, a policy of clear cutting affronts many environmentalists, whose identities are rooted in the preservation of nature. Threats to identities in environmental disputes also contribute to unifying stakeholder frames and positions. For example, in a dispute that centered on Native Americans' scarce water supplies, individuals solidified into a group when they feared the loss of cultural values and community existence (Folk-Williams, 1988). Thus, when disputants feel that their core values are at stake, some parties prefer to fight rather than sacrifice their identities (Rothman, 1997). Threats to identity, then, appear to contribute to entrenchment and escalation of conflict (Northrup, 1989).

In Gray's (1997) model of framing, five types of identity frame characterize different views of self in an environmental dispute: (1) societal role, or the identity one holds in society at large or a community, for example, activist, environmentalist, or victim; (2) ethnic or cultural identity, or the affiliation with a racial or cultural group; (3) place identity, or the alignment of self with a location or physical space; (4) institutional identity, or the view of self as a member of an occupational, professional, or organizational group; and (5) interest identity, or the alignment of self with particular issues, needs, and wants of a stakeholder group. Although individuals may embrace different views of self, they typically perceive or situate one of these frames as primary in accounting for their role in an environmental dispute. These frames may shift over time, indicating that the labels that parties attach to a conflict also shift.

Gray, Jones-Corley, and Hanke (1999) have sought to empirically validate this typology of identity frames by investigating the use of various frame types in a large set of case studies of intractable environmental conflicts. Our first case study reports a qualitative frame analysis of one such environmental dispute: the long-standing battle over the Edwards aquifer in the hill country surrounding San Antonio, Texas. We use this case to illustrate the relationships among group identity, identity frames, and action depicted in figure 13.1.

Case Study I: The Edwards Aquifer Conflict

The Edwards aquifer is a unique underground limestone formation that extends from south of Austin to 100 miles west of San Antonio. As a common-pool resource, it has a finite storage capacity and its water level is highly variable and sensitive to rainfall fluctuations. The aquifer is the primary source of water for a five-county region, including the city of San Antonio, the ninth largest city in the United States.

The key issues in this dispute are (1) the management of a scarce common-pool resource limited by physical structure and used by many interdependent stakeholders, (2) concerns for property rights, (3) the regulation of water allocation and mode of distribution, and (4) the effects of overpumping on the water quality of the aquifer and the five endangered species that live in the aquifer's springs.

Method   This study of identity frames in the Edwards aquifer dispute began with collecting newspaper articles and documents on the conflict from 1988 to 1996. Newspaper articles were located through the NEWS-BANK database, press clippings collected by the Texas Natural Resource Conservation Commission (TNRCC), and contributions from stakeholders interviewed in the study. The search resulted in 122 articles from NEWSBANK, 58 articles from the TNRCC database, and an additional 60 articles from stakeholders, for a total of 240 articles.

In addition, the researchers conducted face-to-face and telephone interviews with sixty-eight stakeholders, including individuals aligned with environmentalists; farmers and irrigators; city and state governmental officials; local, state, and regional water agencies; media personnel; spokespersons from business and industry; and mediators in the dispute. The stakeholders were identified through newspaper stories and a snowball sample in which each stakeholder interviewed recommended other key people to include.

The transcripts of the interviews were coded according to five major identity frames: societal role, ethnic or cultural identity, place identity, institutional identity, and interest-based identity. The institutional and interest-based frames were further divided into positive, negative, and neutral statements based on the adjectives included in the statement. The unit of analysis for coding was a relevant thought unit within a contribution or a sentence. These categories were mutually exclusive but were not exhaustive of all the data in the articles or interview transcripts. Each major category was treated as a unique data set, with different unitizing and reliability calculations. The reliability calculations for the identity category for the interview data were 0.91 (interrater agreement), with a Cohen's kappa of 0.85. Disagreement among coders was resolved through consensus coding. In the interview transcripts, identity statements were roughly 15 percent of the overall contributions coded. (The coding scheme on framing for this study included seven categories: identity, characterization, views of social control, views of nature, conflict management preferences, power, and intractability. Since this chapter centers on identity, the other categories are excluded from this analysis.)

Data Analysis This data analysis is based on the coding of sixty-eight interviews, supplemented by qualitative data from the newspaper articles to illustrate changes in identity framing across four key events: the demise of the regional water development plan (1988–1991), the declaration of the Edwards aquifer as an underground river (1991–1992), the Sierra Club lawsuits (1991–1996), and the legislation creating the Edwards Aquifer Authority (EAA) (1993–1996).

Overall, institutional identity frames dominated the image that stakeholders had of this conflict. Of the 848 total identity statements coded, 49 percent of them related to professional, occupational, and organizational jurisdiction. Of these institutional statements, 64 percent were positive. Statements such as, "Our role is to be proactive, to conserve, and to protect the endangered species, " is an example of a positive institutional identity statement made by a Sierra Club member.

Interest-based identity statements were the next category used frequently with 356 statements, or 41 percent of the identity remarks. As with the institutional frame, 69 percent of the interest-based identity frames were positive. Statements such as, "We have to keep the springs flowing, the water in the wells, and the economy strong" made by a member of the tourism industry illustrate this positive interest-based identity. Only seventy-two identity statements fell into the place frame; hence, it was not a major reference point for identity. For example, one of the farmers remarked, "This land is my whole way of life and it has been so for generations. I will do anything to protect my God given right to this land." In these examples, place merges with self to become one identity in the soil.

Identity frames for the conflict as a whole changed over time. Stakeholder groups, as depicted in the media, formed the level of analysis for tracking the shifting role of identity frames. The early stages of this conflict were characterized by interest-based identity frames, as depicted in statements about defending the sacred principle of "private property rights," preserving the springs, and protecting local economies. For example, a comment by a member of the San Antonio City Council that "the farmers are unyielding in their belief in the property right principle." Or a farmer's remark, "I see it the way I see gun control—you can have my water when you peel my cold, dead hand off my pump." And a comment by the San Antonio mayor that reinforced his group's interest-based frame, "This is a community issue . . . requirements for aquifer pumping limits would cause extraordinary and irreparable harm to the city and its inhabitants (Dilanian, 1996, p. 1A).

These group identities shifted to place-based frames during the early struggles between rural and urban counties with statements such as, "urban areas have a greater need for drought management than do rural communities. The Regional Water Management Plan favors the more populated eastern counties over the rural users in the west" (Wood, 1988, p. 7A). This form of polarization between the urban and rural stakeholders led to identities rooted in ingroup and outgroup membership.

With efforts to declare the Edwards aquifer an underground river, identities shifted to institutional domains, particularly ones vying to develop and implement rules to manage the aquifer. From this point on, the institutional role remained the dominant identity frame—with stakeholders focusing on the salience of their own agencies, constituents, or organizations in this dispute. For example, the Medina County Underground Water Conservation District reacted to the Texas Water Commission by declaring their agency "the only governmental agency authorized by the Texas Legislature to conserve and regulate the use of Edwards water [in their region]" (Winingham, 1992, p. 1C).

The battle of the institutional identities moved into the courts, with regional, state, and federal agencies struggling for control of the aquifer. The environmentalists, led by the Sierra Club, saw no hope for effective regulation without federal intervention and emergency pumping regulations, and they forced the hand of the federal government by filing lawsuits. State agencies argued for stringent rules and "extremely protective regulations," while local and regional water districts in the western counties contended that they had jurisdiction over their own water. Institutional roles, then typically those of watchdog, protector of resources, or ruler, remained pronounced as an identity frame for the latter stages of this conflict.

With the passing of Texas Senate Bill 1477 in 1996, the state legislature created the Edwards Aquifer Authority and gave it the power to regulate and enforce pumping limits. Ironically, as the one agency that had constitutional authority, the EAA did not project a strong institutional identity. Many of the elected EAA board members spoke in the voices of their constituents, characterized by their identity frames rather than those of a central regulatory body. Thus the EAA board members lacked a group identity that was separate from the institutional identities of their constituents.

Although the stakeholders eventually admitted that the EAA was the primary regulatory body, they defined the conflict from their own institutional position. That is, when they were asked to characterize what the conflict was about, they responded with a strong institutional identity: "it is about keeping our factories open, protecting water quality, and complying with regulatory rules." The stakeholders viewed the different jurisdictions of federal, state, regional, and local authorities as central to their individual identities in this conflict.

In effect, group identity played a major role in all stages of the Edwards aquifer dispute. Overall, the group identities of the stakeholders were linked to institutional and interest-based frames, with some recognition of place. Moreover, rather than being a static concept, identity in this case referred to the way parties framed or prioritized their group roles over time. As events changed, the source or type of group identity shifted to enable disputants to invoke actions, justify positions, and respond to the other parties. The farmers and irrigators in the western counties retained a central focus on private property rights and opposed a permanent cap on pumping and any provisions for buying and selling water rights. Even though they emphasized interest, place, and institutional roles as ways of engaging their disputants, they retained their societal identity as landowners and even carried it into the decision processes of the EAA, as evidenced by a vote against emergency pumping limits—one that split the board 7 to 6 along regional lines.

Environmentalists retained strong institutional identities and while they proclaimed the need to protect the endangered species in the springs, they defined their group role through legal and jurisdictional boundaries. San Antonio entered the conflict as the urban bully with a strong place identity. The political battles over water within the city led to a shift in identity frames to institutional roles. Each of these stakeholders relied heavily on their group identities to frame issues in the dispute, select courses of action, and react to various rulings and mandates throughout this intractable dispute.

In summary, the Edwards aquifer case illustrates how environmental conflict can be generated and perpetuated by a clash of existing identities between different groups of stakeholders. Differences in the way the dispute was framed across stakeholder groups contributed to its intractability. Given that variability in identity frames is likely to be the rule rather than the exception in environmental conflicts, what can be done to bridge the divide for diverse parties trapped in an intractable conflict? The dynamic nature of shifts in identity frames over time suggests a possible strategy: create a "safe" environment that encourages stakeholders to "reframe" the situation in terms of a superordinate group identity that rises above the respective individual identities of the parties. The next section presents a second case study in which this strategy was implemented.

Case Study II: The San Antonio Watershed Restoration Council

This case was part of an interdisciplinary research project designed to develop, deliver, and evaluate a new approach to public participation in local watershed restoration efforts. This method was based on the principles of collaborative learning (CL) (Daniels & Walker, 2001). The research site was the San Antonio River Basin in Texas, with a specific focus on two urban stream systems, Salado and Leon Creeks, that pass through the city of San Antonio.

Theoretical Foundations of Collaborative Learning   Collaborative learning is a promising method for structuring and facilitating communication among stakeholders in the context of ecosystem management. The CL process (Daniels & Walker, 2001) is grounded in theoretical work on soft systems methodology (Checkland & Scholes, 1990; Wilson & Morren, 1990) and alternative dispute resolution (Fisher, Ury, & Patton, 1991; Gray, 1989). The concept of "soft systems" is an extension of theoretical work on systems analysis (Senge, 1990) and experiential learning (Kolb, 1986).

The basic assumption in soft systems methodology is that the management of complex problem situations demands a different approach than the typical "hard" systems method used in engineering. Such hard systems methods focus on outcomes instead of processes, and consequently do not attach importance to learning. Management of an ecosystem is a situation where problems are often characterized by a high degree of uncertainty and equivocality. The soft systems approach focuses on situation improvements that can result from active learning and debate.

The CL process also promotes the development and identification of stakeholder concerns and underlying interests. Recent theory and research on negotiation and mediation has adopted an interests-based approach. One viewpoint (e.g., Bazerman & Neale, 1992; Fisher, Ury, & Patton, 1991) maintains that traditional positional approaches to negotiation may be inefficient because the positions taken by parties in conflict are often extreme and obscure the underlying interests (i.e., needs, concerns, values) that parties seek to advance through negotiation. The structured set of CL activities is designed to move participants away from positional strategies and toward the identification of mutual interests and joint gains from collaboration. Moreover, the presence of outside facilitators during CL sessions permits the use of effective principles from mediation theory and research (Gray, 1989).

Methodology of Collaborative Learning Meetings   The CL meetings were designed to enable the stakeholders to move through four phases of application. Each new phase incorporated activities and communication processes that had been developed during the previous phase. First, the participants were informed about the CL process and received training in collaborative discussion and debate. Second, a common knowledge base on the major issues affecting water quality in the watershed was created for them through formal presentations by experts, panel question and answer sessions, and informal small group discussions. Third, active learning exercises were used to help the participants think systemically about the watershed ecosystem and enable them to identify key issues, concerns, and interrelationships among variables affecting water quality. During this phase, the participants began to generate specific suggestions for improving the current situation (i.e., the existing water quality in the watershed) and then share and refine their ideas with other participants through a structured small group discussion process. Fourth, through collaborative debate with other stakeholders, a final set of improvement suggestions was organized, discussed, and refined.

Thirteen CL meetings were held over 14 months, from November 1999 until December 2000. As the participants moved through these phases, we monitored communication patterns to track their sense of personal identity as it related to group identity. The CL sessions were designed to use group identity to broaden the council members' indi-vidual identities. When the participants first began attending meetings, each had a specific point from which they identified with the watershed. As the process unfolded over time, individual identities became enmeshed in the group identity, which was mutually shaped by all participants. The responses of those who developed strong group identity through their participation indicate that while this new identity did not minimize indi-vidual interests, it enabled members to position themselves in relation to the identities of other members. This process encouraged all participants to identify more systemically with the entire watershed, without losing their original sense of personal identity.

Qualitative Analysis   The meetings were both audio- and videorecorded for review by researchers. Over the 14-month period, the stakeholders were asked to participate in a variety of activities. A brief summary of the responses of one watershed council group to these activities illus-trates that as a sense of group identity emerged, council members behaved differently than they had previously.

At the beginning of the first council meeting, the facilitators spread a long length of blue fabric across the floor of the meeting room. They asked the council members to imagine that the fabric was their creek, and to position themselves at an appropriate point along its banks. The members came forward tentatively, and by using the locator markings on the fabric (e.g., points where major highways crossed the creek, parks, and schools), chose a position. The facilitators then invited the council members to introduce themselves with a statement that described their purpose in attending the meeting. A sampling of the self-introductions indicates the diverse identities of group members.

Because he had positioned himself at the uppermost point along the creek, Calvin was the first participant to introduce himself. He began: "I'm with the San Antonio Water System, and I didn't really know where to stand along the creek. I'm actually concerned with all the water quality along Salado Creek. My department is concerned, obviously with water quality. We don't really care where we go; if it's dirty we help clean it up." Calvin illustrates the attitude shared by most agency participants. Although they did not identify with a particular location along the creek, they identified strongly with the aspect of the creek for which they held technical responsibility. In Gray's (1997) typology, Calvin's statement reflects an institutional identity frame.

Ramona, who lives near the creek and is interested in improving the aesthetic quality of the city, said, "I am seriously, anxiously, interested in seeing this creek reach the beautification that I know it can. It can be beautified." Hank, a farmer whose family members have lived on the creek since they emigrated from Italy in the 1890s, told participants, "We live up the road there on ———— farm. It's our livelihood. We farm there; a family farm. We are concerned about the quantity and quality of the water so we can keep the crops growing." Cathy, a science teacher, was "concerned about upstream development, nonpoint source pollution, and about truly educating our children about the tremendous value of the creek." She compared the creek to "an emerald necklace," and said she was participating because she wanted "to preserve the treasure." Daniel, a retired military officer, spoke directly about his concern with flooding: "I'm with the ———— Home Owners Improvement Association. I live north of ———— Rd. and I-35, and when Salado Creek flooded it came right into my house and so I'm definitely interested in what slows down Salado Creek, and I am hoping that it won't happen again." We can see examples of two additional identity frames in these self-introductions: societal role (Hank) and interest-based (Ramona, Cathy, Daniel).

These individuals represent the primary motives that brought people to council meetings. People came because they had a professional responsibility or because they wanted to make the creek beautiful, use water from the creek to make a living, restore ecological integrity to the watershed, or prevent damage from flooding. At this point, they had not imagined the possibility that concerns so diverse could form any unifying motive to act. As they engaged in joint learning experiences, however, a unifying motive emerged, and they developed a set of technically feasible recommendations for improving the quality of their watershed that incorporated all of the concerns mentioned here.

Thinking that the first meeting was crucial to setting a positive tone, we (the facilitators) asked the mayor to attend and worked with him to develop an appropriate set of remarks. Following the self-introductions, he expressed his enthusiastic support for the CL process, and his hope that the council would become a positive force for change in the city's management of the watershed. He encouraged its members to think systemically about the watershed, and to include concerns from all reaches of the watershed in their recommendations. When he told them that the city would be able to make funds available to assist in implementation of council recommendations, Daniel immediately raised his hand and requested that the funds be spent to clean out the creek bed immediately upstream of, and throughout his neighborhood. He asked that all brush and trees be removed from the creek banks, and that the creek bed be lined with concrete. The mayor responded by saying he hoped Daniel would think about the entire watershed, and would consider the needs of other council members before deciding on such a drastic course.

The citizens met monthly (in 4-hour workshops) for approximately a year. Meetings included the introduction of new ecological and socioeconomic and political information about the watershed. The members were then invited to devote part of the time to discussions of how this information could be applied to improving the current situation. During the first four meetings, individuals simply reiterated their own motives during discussions. Gradually, however, they began to integrate their own needs into what they perceived to be the needs of fellow council members. As the group progressed, it became increasingly clear that participation in council meetings was central to developing an increasing desire and competence to participate in environmental management decisions. At the eighth meeting, one small group, including members representing all the interests mentioned during the first meeting, reported its discussion to the council as follows:

Well, we wanted to protect the ecology and water quality. We felt like work that needed to be done, in this area that was flooded, that we have to consider protection of property and life and that would be the driving force in that area. And what we would say is that would be the driving force, but also keep in mind protecting ecology, protecting water quality and protecting wildlife in the area. And at the same time develop recreation facilities to serve people in that area and also all citizens so that the whole city would benefit which would encourage spending money on the creeks. So that is what we were looking at, that protection of life and property would be in the forefront, but flooding improvements should be done in such a way to protect the wildlife and also provide recreation opportunities and other economic opportunities.

From this point, the council members spent an increasing portion of each meeting developing and refining a set of recommendations to present to the city. They also began discussing how to encourage implementation and enforcement of their recommendations.

Ironically, the council's hard work was nearly derailed when local citizens voted in favor of a tax to provide additional funds for watershed restoration. Extensive publicity for the new funding drew the attention of many new participants. Suddenly, at a meeting at which participants had expected to make final revisions to their recommendations for the mayor, attendance doubled. Participants who had attended meetings for a year glanced uncomfortably at each other as new attendees vented their concerns, making such statements as:

I know what I own and ain't no one gonna mess with it. I turned 88 last month, and I'm gonna be out there another 50 years cause Genesis 6 and 3 gives me 120 years. On top of that, you don't want to deal with my relatives. I got 50 kids , I had five, and they went out and brought home five. And that gives me ten. They gave me twenty grandkids and they gave me twenty great-grandkids. And I got a man sitting here that is the grandfather of two of my great-grandkids. So you don't want to deal with them. You better deal with my wife of 57 years and me on buying that water rights and buying that land, because other people won that land up and down that Salado and you ain't just gonna walk in and say we're gonna take it. Because a lot of them are meaner than me, and I'm as mean as a junkyard dog.

Because we had essentially turned management of the group over to local participants, we did not intervene. After listening to similar inflammatory comments for approximately 30 minutes, Cathy, one of the seasoned participants, attempted to redirect the conversation. Other "veteran" group members, recognizing that the behavior of the new attendees was threatening the attributed consensus about how to improve the watershed, as well as how to behave as a council member, soon joined her. They had developed these informal norms through months of difficult interaction and were not prepared to give them up lightly.

Cathy started by explaining the CL process they were using. She added that the facilitators "are here to help us make our own decisions about the creek, not to tell us what to do." She then introduced a group of people designated as "the nominating committee" (i.e., participants who, at the request of the group, had spent 2 months developing a slate of nominees willing to serve as officers in a continuing council). Hank, John, Janet, and Tina each spoke about the work the council had done, and its future direction. Janet summarized the concerns of council members when she explained, "we are concerned that we need a place at the table to spend this money. We are concerned that people will come out from under rocks to spend the money on things we don't want. We need to protect the trees. We need to help people and we need safety and protection. We need to institutionalize ourselves." Janet's presentation reminded everyone in the room that the council had developed a deep sense of identification with the entire watershed, and that this understanding was the result of hard work. It was not something to give up lightly.

Following the nominating committee's presentation, Cathy took charge of the meeting. Each time a new attendee attempted to return the conversation to a single interest at the expense of the group identity, she turned to an experienced member to help move the discussion back to the goal of watershed improvement. At the next meeting, the attendance pattern changed again. The seasoned participants returned, along with only one quarter of the new attendees. The new attendees, for whom this was a second meeting, began accepting the norms established over the past year and joined in planning an upcoming field trip to further strengthen the sense of identity with the entire watershed.

We left the group in December 2000 and the council has since developed into a private foundation. Electronic messages received from members indicate its continuing participation in city governance issues relevant to the watershed.

Conclusions

We have presented here two case studies to explore the role of group identity in mediating between knowledge and action. What conclusions can be drawn from these cases that inform the conceptual framework in figure 13.1?

The Edwards aquifer case reveals how group identities can shape the positions and actions taken by diverse stakeholder groups and why the conflict escalated over time into an intractable situation. This case addresses the right-hand portion of figure 13.1, namely, the links among group identity, identity frames, and action. For example, the imposition of a superordinate authority (the EAA) to manage the Edwards aquifer threatened the identities of several stakeholder groups, most prominently the farmers and irrigators.

The frame analysis suggests a shift from interest-based and place identities toward institutional identities as the conflict moved into the lawsuit phase and the contentious struggles over the legitimacy of the EAA. This frame shift makes sense because the institutional identity frame is more functional for stakeholder groups when they are faced with preparing for legal court battles.

Thus there is some evidence in this case that group identity, identity frames, and stakeholder action are linked reciprocally. Moreover, this case analysis reveals the dynamic aspect of identity frames and how external events (i.e., actions by stakeholder groups) can influence (or be influenced by) the type of identity frames adopted by stakeholders over the course of this multiparty conflict. It also suggests that certain types of identity frames (e.g., institutional) may be more likely to lock stakeholder groups into positions that further escalate the conflict. In contrast, interest-based identity frames may be more amenable to negotiation and mediation efforts.

The San Antonio Watershed Council case suggests how a new group identity can be created through the structured communication process of collaborative learning. The results of this case are relevant to the left-hand portion of figure 13.1, showing the process by which knowledge leads to the development of group identity. The case also illustrates how the relationship between group identity and action is mediated by normative influence.

The three hypothesized antecedents to formation of group identity—consensus attribution, group interaction, and generative process—are all present in the qualitative account of the group's history. The fact that council members gradually achieved a systemic orientation toward their watershed is an example of the group's attributed consensus. This was reached only through extensive group interaction over time. The CL process was designed to be iterative, with new information presented in stages to council participants and with adequate time allowed for the participants to develop new understandings of their watershed ecosystem.

Another important observation is the power of the council's group identity in changing its members' behavior patterns during later meetings. While members arrived at the beginning of the workshops with very different agendas and personal identities, these differences were accommodated by the end of the meeting series, and a new council identity prevailed, despite direct challenges from new members. A normative influence was apparent as experienced council members intervened to ensure that norm violations by new members did not disrupt the attributed consensus that developed through intensive and sometimes difficult interaction over the 14-month CL process.

In summary, the main finding from the San Antonio Watershed Council case is that repeated communication opportunities among stakeholders, structured by collaborative learning principles, allowed participants to develop a new, superordinate group identity that permitted the group to act. The group's final product, a consensus-based set of council recommendations to improve the quality of life in their watershed, would have been unlikely without the support of their newly forged group identity. Although we do not claim that this emergent group identity was the only force guiding the participants' behavior, we do conclude it was most likely a significant contributing factor.

Theoretical Implications

There are two significant theoretical implications of these conclusions. First, the qualitative frame analysis of the Edwards aquifer case supports Gray's (1997) assertion that identity plays a significant role in the genesis and escalation of intractable environmental conflicts. This suggests that further elaboration of the framing approach is warranted.

Second, the qualitative results from the San Antonio Watershed Council case are consistent with experimental studies in social psychology that demonstrate the causal role of group identity in enhancing cooperation in social dilemmas (e.g., Brewer & Kramer, 1986; Kramer & Brewer, 1984; Orbell, van de Kragt, & Dawes, 1988). The current debate in this literature concerns isolating the "best" explanation for why group discussion increases cooperation in social dilemmas (see Ostrom, 1998; Sally, 1995, for reviews). Orbell et al. (1988) proposed that group identity is the most likely cause of this effect. While other theoretical accounts have been offered (e.g., Kerr and Kaufman-Gilleland, 1994; Bouas & Komorita, 1996), there is consensus that group identity is a contributing causal factor in explaining this effect of discussion. The convergence of results from controlled laboratory experiments and our field study builds confidence in the conclusion that the collaborative learning communication process helped promote group identity among council members.

Implications for Practice

In terms of practice, the San Antonio Watershed Council case suggests that collaborative learning can be an effective method for promoting constructive dialogue and managing conflicts among diverse stakeholders in a water resource management context. Some caveats to this generalization are in order, however. Collaborative learning (Daniels & Walker, 2001) is not a panacea for all problems involving stakeholder conflict. Daniels and Walker (2001) emphasize the need for a thorough situation assessment prior to using a CL intervention. They outline specific conditions under which CL processes are most likely to be successful.

In our particular situation, we employed CL because we viewed the context surrounding the group as favorable. The initial, overt conflict level was mild to moderate; stakeholders had not yet polarized into rival subgroups; and the potential for a real, positive change resulting from collaborative work by the council was moderate to high. Thus the positive experience observed in our case may not generalize to other conflict situations. We suggest that collaborative learning can be a valuable technique for managing stakeholder conflict when it is used wisely in the right situations.

References

Ashforth, B. E., & Mael, F. (1989). Social identity theory and the organization. *Academy of Management Review, 14,* 20–39.

Bazerman, M. H., & Neale, M. A. (1992). *Negotiating rationally.* New York: Free Press.

Brewer, M. B., & Kramer, R. M. (1986). Choice behavior in social dilemmas: Effects of social identity, group size, and decision framing. *Journal of Personality and Social Psychology, 50,* 543–549.

Bouas, K. S., & Komorita, S. S. (1996). Group discussion and cooperation in social dilemmas. *Personality and Social Psychology Bulletin, 22,* 1144–1150.

Carbaugh, D. (1996). *Situating selves: The communication of social identities in American scenes.* Albany: State University of New York Press.

Checkland, P., & Scholes, J. (1990). *Soft systems methodology in action.* New York: Wiley.

Daniels, S. E., & Walker, G. B. (2001). *Working through environmental conflict: The collaborative learning approach.* Westport, Conn.: Praeger.

Dilanian, K. (1996). Thornton files Bunton appeal. *San Antonio Express-News,* Aug. 27, pp. 1A–6A.

Douglas, M., & Wildavsky, A. (1982). *Risk and culture: An essay on the selection of technological and environmental dangers.* Berkeley, Calif: University of California Press.

Fisher, R., Ury, W., & Patton, B. (1991). *Getting to YES* (2nd ed.). New York: Penguin.

Folk-Williams, J. A. (1988). The use of negotiated agreements to resolve water disputes involving Indian rights. *Natural Resources Journal, 28,* 63–103.

Gray, B. (1989). *Collaborating: Finding common ground for multiparty problems.* San Francisco: Jossey-Bass.

Gray, B. (1997). Framing and reframing of intractable environmental disputes. In R. J. Lewicki, B. Sheppard, & B. Bies (Eds.), *Research on negotiation in organizations* (Vol. 6, pp. 163–188). Greenwich, Conn.: JAI Press.

Gray, B., Jones-Corley, J., & Hanke, R. (1999). The framing of identity, rights, and values in environmental disputes. In R. J. Lewicki (Chair), The framing of intractable environmental disputes. Symposium conducted at the Annual Meeting of the Academy of Management.

Hoare, C. H. (1994). Psychosocial identity development in United States society: Its role in fostering exclusion of other cultures. In E. P. Salett & D. R. Koslow (Eds.), *Race, ethnicity, and self: Identity in multicultural perspective* (pp. 24–41). Washington, D.C.: National Multicultural Institute.

Hogg, M. A., Terry, D. J., & White, K. M. (1995). A tale of two theories: A critical comparison of identity theory with social identity theory. *Social Psychology Quarterly, 58,* 255–269.

Kerr, N. L., & Kaufman-Gilleland, C. M. (1994). Communication, commitment, and cooperation in social dilemmas. *Journal of Personality and Social Psychology, 66,* 513–529.

Kolb, D. A. (1986). *Experiential learning: Experience as the source of learning and development.* Englewood Cliffs, N.J.: Prentice-Hall.

Kramer, R. M., & Brewer, M. B. (1984). Effects of group identity on resource use in a simulated commons dilemma. *Journal of Personality and Social Psychology, 46,* 1044–1057.

Northrup, T. A. (1989). The dynamic of identity in personal and social conflict. In L. Kriesberg, T. A. Northrup, & S. J. Thorson (Eds.), *Intractable conflicts and their transformation* (pp. 55–82). Syracuse, N.Y.: Syracuse University Press.

Orbell, J. M., van de Kragt, A. J. C., & Dawes, R. M. (1988). Explaining discussion-induced cooperation. *Journal of Personality and Social Psychology, 54,* 811–819.

Ostrom, E. (1998). A behavioral approach to the rational choice theory of collective action. *American Political Science Review, 92*(1), 1–22.

Prelli, L. J. (1989). *A rhetoric of science: Inventing scientific discourse*. Columbia: University of South Carolina Press.

Putnam, L. L., & Holmer, M. (1992). Framing, reframing, and issue development. In L. L. Putnam & M. E. Roloff (Eds.), *Communication and negotiation* (pp. 128–155). Newbury Park, Calif.: Sage.

Roland, A. (1994). Identity, self, and individualism in a multicultural perspective. In E. P. Salett & D. R. Koslow (Eds.), *Race, ethnicity, and self: Identity in multicultural perspective* (pp. 41–59). Washington, DC: National Multicultural Institute.

Rothman, J. (1997). *Resolving identity-based conflict in nations, organizations, and communities*. San Francisco: Jossey-Bass.

Sally, D. (1995). Conversation and cooperation in social dilemmas: A meta-analysis of experiments from 1958–1992. *Rationality and Society, 7*, 58–92.

Scheff, T. J. (1967). Toward a sociological model of consensus. *American Sociological Review, 32*, 32–46.

Senge, P. M. (1990). *The fifth discipline: The art and practice of the learning organization*. New York: Currency Doubleday.

Tajfel, H., & Turner, J. C. (1985). The social identity theory of intergroup behavior. In S. Worchel & W. G. Austin (Eds.), *Psychology of intergroup relations* (pp. 7–24). Chicago: Nelson-Hall.

Taylor, D. E. (2000). The rise of the environmental justice paradigm: Injustice framing and the social construction of environmental discourses. *American Behavioral Scientist, 43*(4), 508–580.

Taylor, D. M., & Moghaddam, F. (1994). *Theories of intergroup relations: International social psychological perspectives* (2nd ed.). Westport, Conn.: Praeger.

Tversky, A., & Kahneman, D. (1974). Judgment under uncertainty: Heuristics and biases. *Science, 185*, 1124–1131.

Wilson, K., & Morren, G. E. B. (1990). *Systems approaches for improvements in agriculture and resource management*. New York: Macmillan.

Winingham, R. (1992). Counties request water districts; Farm bureau set to deliver petitions for Kinney and Uvalde. *San Antonio Express-News*, Nov. 25, p. 1C.

Wood, J. (1988). EUWD, city prepares water rationing plea. *San Antonio Express-News*, June 25, p. 7A.

14

Constructing and Maintaining Ecological Identities: The Strategies of Deep Ecologists

Steve Zavestoski

How is it that so many people claim to be concerned about the environment while at the same time making life-style choices that lead to environmental destruction? This question has plagued me since I began identifying with the deep ecology movement. In the middle of my undergraduate studies at the time, I was also beginning to develop an academic identity, one rooted in the social-psychological tradition of symbolic interactionism. Combining these two identities, I began focusing on the significance of ecological identities in answering my question. In the same way that various types of social identities shape, guide, and make meaningful social interaction, I thought, ecological identities may be essential in shaping, guiding, and making meaningful human interaction with the environment.

This view led me to the conclusion shared by other social scientists that environmental problems are more accurately understood as problems of social organization (Bell, 1998). Social organization, which shapes our experience of reality, helps make our social lives meaningful. However, our socially constructed reality, when informed by weak or nonexistent ecological identities, inadequately accounts for our impacts on the natural environment. If environmental problems are seen as problems of social organization, the organization of the self becomes significant.

Because we rely on others to attach the same meanings to our identities that we do, our identities allow us to anticipate how others will react to us. Why do so many people report concern for the environment yet fail to make decisions to minimize their impact on the environment? Perhaps this concern is not linked to an ecological identity, and therefore we fail to consider others' reactions to our environmentally destructive behavior, much less the reactions of environmental "others."

There is scant research into the significance of the self or identity in orienting human relationships with the environment. On the other hand, the relationships among environmental beliefs, attitudes, and behaviors have been studied at length (Gardner & Stern 1996; Stern & Dietz, 1994; Stern, Dietz, Kalof & Guagnano, 1995). The extension of this research to an emphasis on the self seems quite logical. If values and attitudes influence people's behavior toward the environment (Axelrod, 1994; Axelrod & Lehman, 1993; Dunlap, Grieneeks, & Rokeach 1983; Gamba & Oskamp, 1994; Gutierrez Karp, 1996; Hines, Hungerford, & Tomera, 1986; Schultz & Oskamp, 1996; Schultz, Oskamp, & Mainieri, 1995; Stern et al., 1995), then the self-concept, which is the cognitive structure in which values and attitudes are mapped, should also influence this behavior.

I first define the terms *self*, *self-concept*, and *identity*, in order to move toward a conceptualization of ecological identity. An ecological identity, I argue, gives an individual the ability to connect her or his social behavior to its environmental impacts. I then draw on the experiences of those in the deep ecology movement to illustrate the challenges of maintaining an ecological identity in a society that is primarily organized around exclusively social identities.

Self, Self-Concept, and Identity

As we engage in social interaction we learn from other's reactions to our behavior what responses to expect in future social situations (Mead, 1934). Mead refers to the set of anticipated responses that individuals form as the "generalized other."

Drawing on the generalized other allows a dialogue to take place between the "I," the spontaneous part of the self, and the "me," or reflective part of the self. It is this internal dialogue that is the self. The self-concept is the product of this process.

For further conceptual clarity, identity theorists define the self-concept in terms of a set of identities (Foote, 1951; Stone, 1962; Rosenberg, 1979; Stryker, 1980, 1991). For Stryker, these identities are hierarchically arranged within the self-concept. An identity's salience, or the likelihood it will be drawn upon in a social situation, is a function of the individual's commitment to the identity. Identity salience and commitment represent an interactive complex from which, given the values of each variable, behavior can theoretically be predicted.

The self-concept, then, is the sum of all the individual's thoughts about her or himself as an object (Rosenberg, 1979). Identities allow us to categorize these thoughts according to our specific roles in society. Identities organize the content of the generalized other, and relate individuals to particular social positions (McCall & Simmons, 1966). As Gecas and Burke explain, "In this way the multifaceted nature of the self (each facet being an identity) is tied to the multifaceted nature of society" (1995, p. 45).

Whence Ecological Identities?

If the self-concept is shaped by social interaction, and if identities link individuals to a role in society, what exactly is an "ecological identity?" And how can ecological identities emerge unless we can interact socially with aspects of the natural world? I conceptualize ecological identity as that part of the self that allows individuals to anticipate the reactions of the environment to their behavior.

The deep ecology philosophy, which advocates a broadening of the self to include nature (Naess, 1989), helps to illustrate ecological identities. John Seed's description of his motivation for protecting the rainforest captures such a broadening of the self: "I try to remember that it's not me, John Seed, trying to protect the rainforest. Rather, I am part of the rainforest protecting itself. I am that part of the rainforest recently emerged into human thinking" (Seed, Macy, Fleming, & Naess, 1988, p. 36). Zimmerman's understanding of deep ecology argues that "intellectual conclusions alone are not sufficient to bring about a basic shift in one's attitude toward nature. . . . Such a shift requires a change of consciousness, an intuitive sense of identification with all things" (1993, p. 199).

Although deep ecologists such as Naess speak of an ecological self, others have referred to an ecological identity (Thomashow, 1995). For Thomashow, "ecological identity refers to all the different ways people construe themselves in relationship to the earth as manifested in personality, values, actions, and sense of self;" which results in "nature becom[ing] an object of identification" (1995, p. 3).

Social-Psychological Bases for Ecological Identities

How can deep ecology's understanding of an ecological self or identity be reconciled with the social-psychological account of the self in which the self-concept and identities are defined by social interaction? William James's (1890) conception of the self reveals that in addition to a broad range of social objects, the self can also include one's body and psychic powers, one's clothes and house, one's ancestors and friends, and even one's land. Mead's generalized other is similarly capable of including a broad range of entities:

> It is possible for inanimate objects . . . to form parts of the generalized . . . other for any given human individual, in so far as he responds to such objects socially or in a social fashion. . . . Any object or set of objects, whether animate or inanimate, human or animal, or merely physical—toward which he . . . responds, socially, is an element in what for him is the generalized other. (Mead, 1934, p. 154n)

More recently, Weigert (1991, 1997) has used a symbolic interactionist perspective to propose that interaction between symbol users, such as humans, and nonsymbol users, such as the natural environment, leads to the construction of "environmental others." These theoretical bases for ecological identities are bolstered by an emerging body of research that empirically demonstrates the presence of ecological identities in certain individuals (Bragg, 1996; Statham, 1995).

Ecological identities, at least in part, seem to emerge from direct experiences in nature that reframe individuals' experiences of themselves in light of a connection to a natural world that is exogenous to culture or society. This view is consistent with the social-psychological perspectives described earlier in which the self can contain nonsocial components. Yet as I point out later, while ecological identities may not necessarily be products of social interaction, their existence ultimately requires that social actors react toward people with ecological identities in meaningful ways.

Ecological Identities as Social Identities

Our identities are made meaningful through the reactions others have to our behaviors. So although individuals can have identities grounded in nonsocial experiences—believers in unidentified flying objects, for example—such identities will wither unless there are other social actors who treat the identities as meaningful.

An ecological identity may compel us to bicycle to work, but constant assaults by automobile drivers who see bicyclists as a nuisance are not the types of reactions that will sustain such an identity. And because the incremental impact our decision has on lessening global warming is not detectable by even the most finely tuned scientific instruments, much less our own senses, we cannot expect a meaningful reaction from nature either. Therefore, because the environment does not respond in socially meaningful ways to our actions, we depend on the responses of social others to validate the actions guided by our ecological identities.

What significance, then, do ecological identities hold? When the meanings of identities are shared by the actors in a social setting, the actors are able to anticipate how each of them will act, and thereby coordinate actions in a way that maintains social order. My students and I have shared understandings of the meanings of my identity as a professor and their identities as students. This shared understanding may be disrupted, however, if I arrive at class sweaty, out of breath, and with a grease stain on my pants leg as a result of having ridden my bike to class. If my students are aware of my ecological identity, and understand that it takes precedence over my identity as a professor, they may understand my appearance while retaining the other behavioral expectations they have of their professors. If they are not aware of my ecological identity, I may lose their respect or threaten my own self-esteem as I deal with their giggles or whispering behind my back. Put simply, an ecological identity's significance, and its potential to result in environmentally sustaining behavior, rests in its link to social identities.

If ecological identities rely on the social meanings that get attached to our actions, constructing ecological identities becomes quite a challenge. Environmentally destructive behaviors such as driving automobiles and using air conditioning tend to carry positive social meanings, not negative ecological meanings. Among most Americans, for example, sport utility vehicles make statements about the driver's life-style rather than about his or her contribution to global warming. Where does one turn to learn the ecological meanings of these behaviors and to get positive social feedback when acting on the ecological identities these meanings inform? My own struggle to negotiate the meaning of my ecological identity led me to the deep ecology movement. Ecological identities must be common, I conjectured, among people whose philosophy embraces the idea of expanding our sense of self to include nature. Deep ecologists, therefore, appeared to be a population that could provide insight into the process of constructing and maintaining ecological identities.

Deep Ecologists and Ecological Identities

Identifying with the deep ecology movement provides both the social meanings of the environmental consequences of one's behavior, and the social support needed to maintain one's ecological identity. Through participation in the deep ecology community, for example, I might learn that others have experienced worry or anxiety over the unusual number of warm days we had during the past winter. Such a realization changes my view of my fossil fuel-burning activities. Even though I may have understood that using an automobile contributes to global warming, I have no personal experience of its effects. My exposure to others' distress over my use of a car provides a more meaningful basis for changing my behavior.

Given this, I wanted to explore how ecological identities function within a community of deep ecologists. Given my own struggle to find support for my ecological identity, I also sought to learn more about others' attempts to maintain their ecological identities.

I first observed the strategies that deep ecologists use to construct and maintain their ecological identities during my participation in a 2-week residential workshop sponsored by the Institute for Deep Ecology. As my ecological and academic identities continued to interweave, I realized I could make more systematic observations of deep ecologists. So when I attended the workshop again 2 years later, I conducted interviews and administered a questionnaire measuring the self-concepts of the participants.

Having developed a supportive relationship with the organizers of the workshop, I was introduced as a deep ecologist who was also an academic studying ecological identity. As a result, the attendees saw me more as a fellow participant than as a researcher. I attempted to respect this relationship by conducting interviews informally. I did not tape the interview or take notes, but instead made extensive notes following each encounter. Drawing on my understanding of the deep ecology movement and my interest in ecological identities, I asked questions that prompted the interviewees to explain their relationship to nature, the origins of this relationship, and how others responded to it.

Creating Safe Social Spaces for Ecological Identities

Set at a retreat center in the Puget Sound region of Washington State, the deep ecology workshop began with the participants being welcomed by workshop staff and then sent to campsites and cabins to set up their living spaces. After the staff and instructors were introduced, the participants were separated into groups named after local flora and fauna. These groups would serve as the participants' families and support within the larger group of participants.

Each morning began with optional nature walks, yoga and meditation sessions, and other activities. Silence was observed during breakfast, while lunches and dinners were more social affairs. All the meals were vegetarian. Evenings offered varied activities ranging from singing and storytelling to videos. The bulk of the workshop consisted of plenary sessions in the morning and afternoon, and elective sessions in between.

These sessions included substantive topics such as ecopsychology and strategies for activism, as well as process-oriented activities designed to help the participants further connect to the environment and each other. One session, for example, involved guided meditation with a partner. An instructor narrated the story of the evolution of life on Earth while participants faced their partners with closed eyes. As the narration arrived at the origins of humans on the planet, the participants opened their eyes to connect with their partner. A process of reflecting on the experience and discussing feelings of connectedness followed.

Sessions typically began with either a prayer or some symbolic gesture of thanks to the Earth. Each session also stressed that the workshop was a safe place for the expression of one's desires, fears, passions, and other feelings about the Earth. One session explicitly addressed the participants' feelings of despair in the face of the enormous challenge of transforming the human relationship to the environment, and offered strategies for coping with such feelings.

In another session, a "Council of All Beings," the participants selected an animal or feature of nature to represent. Speaking as a tree, the ocean, or an animal, for example, the participants bore witness before the council of the harm being done to them and the emotional anguish this harm causes. Attendees at the workshop who participated in this role-playing exercise spoke of its transformative effect. Identifying with a part of nature on such a deep level reveals the power of role playing as a way of learning to anticipate responses to our actions, a fundamental function of identities.

Throughout the sessions I attended, the participants repeatedly expressed a concern that their feelings of oneness with nature were not welcome in the "real world" outside the workshop. This concern reflected the fact that in their day-to-day lives, the ecological identities that were being nurtured at the workshop were not responded to meaningfully by others.

Characteristics of Workshop Participants

Through interviews and observation, it became apparent that most of the participants had either special places in nature, a place that had been special to them but was developed or destroyed, or a particular experience in nature that was significant in developing their concern for nature. In interviews the participants explained how expressing their concern for these special places as an emotional attachment or sense of oneness often resulted in strange looks or dismissive reactions.

The workshop participants had chosen to cope with such responses rather than abandon their feelings. A look at some of their demographic characteristics might offer some insight into what types of individuals make such a choice. The majority of the participants were female. Of the fifty surveyed at the second workshop, ages ranged from 18 to 65, centering around 40 years. More than 90 percent had graduated from college. Most of the participants found out about the workshop through informal networks with other environmentalists. The reasons they gave for attending included connecting with other people who shared their feelings for the Earth, learning ways of coping with despair about the state of the environment, and learning techniques for spreading their concern for the environment.

Measuring the Self-Concept

Approaches to measuring the self-concept tend to focus more on self-esteem than the content of the self-concept, and they typically offer respondents choices from a researcher-defined list of identities. Feeling that ecological identities could be manifested in any number of ways, I chose a method that allowed the respondents an unstructured format for expressing their self-concepts. Using the Twenty Statements Test (TST) (Kuhn & McPartland, 1954), which I modified to ten statements, I asked the respondents to respond ten times to the question, as if asking themselves, "Who am I?" Responses to the TST—such as "I am a mother," "I am a doctor," "I am compassionate"—were coded into categories based on Gordon's (1968) scheme for coding the TST.

I also added the category "ecological identity" to capture those responses that reflected a respondent's concern for the environment and awareness of her or his ecological context. These included responses such as "I am an Earth lover," "I am someone who tries to walk lightly on the Earth," "I am a part of nature," and "I am a global citizen." Sixty-three percent of all respondents gave at least one answer coded as ecological identity, and many gave two or more. This was exceeded only by the 83 percent of respondents who gave a response coded "sense of altruism/compassion."

The TST also asked the respondents to rank their identities according to salience (i.e., the identities they most frequently occupy). Ecological identities ranked behind occupational, kinship, altruistic or compassionate, moral, and taste or interest identities. Despite the prevalence of ecological identities in their self-concepts, it appears that the participants did not draw upon these identities as frequently as others.

Maintenance of Ecological Identities

Why would it be that deep ecologists do not draw upon these identities more often? I propose that these identities fail to elicit the feedback from other social actors that is necessary to maintain them. The vast majority of other social actors do not hold or even understand ecological identities. For these persons, an interaction with someone who has developed and is acting on an ecological identity is likely to result in confusion.

This explanation makes sense in light of the fact that identities tend to be strengthened and increase in salience as interactions with others affirm the identity (Burke, 1991; Freese & Burke, 1994). If many of the day-to-day interactions of the participants at the deep ecology workshops are with persons without ecological identities, it is likely that maintaining the salience of their ecological identities is a difficult task. I organize the remainder of my analysis around the strategies that the workshop participants employed to nurture and sustain their ecological identities.

Work

Many participants sought jobs in careers where ecological identities are accepted, or even expected. By taking their ecological identities and weaving them into their occupational identities, some participants seemed to be able to receive daily validation of their ecological identities. One participant described an experience she had during her work as an ecologist: "At a recent workshop I was giving on stream ecology I was moving, flowing with the stream and coming together with other beings in and around the water." In this case, not only does this participant have her ecological identity affirmed through her work, but she also gets to encourage the development of such identities in others.

Many participants worked in fields ranging from environmental activists and environmental consultants, to environmental educators, naturalists, and earth scientists. Many others worked in helping professions as therapists, nurses, educators, and community outreach and nonprofit coordinators. Very few worked in traditional professional jobs. The prevalence of helping careers suggests that even if a profession that explicitly acknowledges an ecological identity is not a possibility, persons with ecological identities will seek careers in which they are able to express their compassion for human others, if not environmental others. Some might contend that what were categorized as ecological identities were, for those working in environmental fields, expressions of their occupational identities. However, through interviews with those participants employed in environmentally related professions, it appears that many participants, driven by their ecological identities, moved from one unfulfilling job to another until they found their current job in which their ecological identities could flourish.

Religion

Practicing a form of Earth-based spirituality is another tactic for nurturing and sustaining ecological identities. The workshop participants often described their various spiritual practices as they related to their environmental concerns. While Earth-based spirituality or paganism were the most popular religions, other participants practiced Buddhism, Taoism, and other eastern religions that have a more explicit Earth ethic than Christian religions. Very few participants practiced traditional religions such as Judaism, Catholicism, or Protestantism. As with their occupations, their choices of religion were aimed at creating lives in which other social actors shared, or at least understood, their ecological identities. Because religion is often the source of values, which most likely cut across the self-concept, a religious practice that does not embrace an individual's ecological identity would lead to extensive internal conflicts.

Social Networks

In addition to their occupational and religious choices, the participants at the deep ecology workshops explored other ways of nurturing and sustaining their ecological identities, including seeking out events such as the workshops, ending such gatherings with a ritual intended to guide the participants back into their daily lives, maintaining communication networks with other participants following the workshops, and in several cases entering into intimate relationships with fellow participants at the workshops.

Many of the participants already had highly developed networks through which they learned of the workshops. Through these networks, they were able to seek out events at which others with similar beliefs, concerns, and identities would be present. This sort of networking was carried on during the workshops through spontaneous announcements by participants of similar gatherings in their own areas, and by the exchange of a directory of names and addresses of all the participants. Seeking out others with shared identities even went beyond an interest in participating in events like the deep ecology workshops. One female participant noted that many of the women she had met at the workshop had left marriages in which their partners did not share their concern for the environment, and had come to the workshop with hopes of finding new, more compatible partners.

Strategies for Returning to the "Real World"

The participants almost unanimously expressed a sense of being in a different world during the 2-week workshops. They described their feelings while at the workshops as "safe," "happy," "at peace," and "contented." But they also recognized the sheltered nature of the workshop and the reality of having to return at the end of the workshop to a world unsupportive of their ecological identities. A group of participants at one workshop decided to deal with this by organizing a closing ritual that would empower people so that the comfort and security of the workshop could be extended.

A final tactic for sustaining ecological identities emerged out of the networking that continued after the workshops. Past participants realized it would be useful to be in touch with and get to know present and future participants. As a result, it was decided that a newsletter would be sent to all past and future workshop participants. The articles included accounts of experiences in which past participants had attempted to express to others their concern for the environment, or passionate accounts of feelings of oneness with nature. Many of the newsletter articles were oriented toward helping readers develop strategies for sharing their level of concern for the environment with others, and for dealing with rejection when those feelings were ignored or disparaged.

An Internet discussion list was another form of networking that grew out of one workshop. Although the discussion was open to anyone, many of the original subscribers to the list were past workshop participants. Members of the discussion group recounted the new perspectives they had gained from experiences with nature and expressed their struggles with trying to get others to understand this new perspective. One member reported the following about her experience at a workshop:

There were . . . many informal opportunities to share . . . heartfelt thoughts, emotions, prayers . . . and rituals. . . . The highlight of the time was the singing and other music we did together. . . . The depth of the vocal harmonies that arose from this group seemed to me to connect with the cosmos, and I usually left those gatherings with a cosmic buzz inside and the feeling that we had been in tune with something much greater than ourselves.

By sharing these sorts of experiences, seeking opportunities to interact with others sharing ecological identities, and using ritual to frame the importance of their feelings and beliefs, the participants at the workshops found ways to nurture and sustain their ecological identities.

Whether intuitively, or out of a sense of survival, it is likely that many of the participants realized the need to surround themselves with likeminded, or like- "identitied," individuals. The workshops saw many late-night conversations, the development of intimate friendships, and many sad good-byes at the close of the meetings. Those at the workshops who either had fully developed ecological identities, or who were nurturing newly discovered ecological identities, thrived in an environment where their beliefs, actions, and expectations were met and responded to in an affirming manner. Conversely, in everyday life, drawing on ecological identities results in misunderstandings or even dismissiveness from those without such identities.

Why Do Ecological Identities Matter?

If individuals consciously draw on their ecological identities, what implications will this have for the environmental impacts of their behavior? Because social actors typically share understandings of social identities, failing to carry out the roles associated with an identity runs the risk of negative reactions from others. The meaning an identity has for us gives us guidelines for how someone with such an identity acts. Ecological identities are no exception to this rule. They *are* exceptional, however, in that their meanings are not widely shared by other social actors, which in turn means that ecological identities are not often affirmed in social interaction.

It makes sense, therefore, that individuals with ecological identities would seek out social contexts where other social actors have a shared understanding of their identities. The deep ecology philosophy (Devall & Sessions, 1985; Devall, 1988, 1993), in fact, maintains that ecological identities have implications not only for the day-to-day decisions people make, but also for more serious decisions such as career choices and political actions. Even for Thomashow, who sees the origins of ecological identities as residing in direct experiences of nature, a widened sense of self results in people "choosing organizations to belong to, selecting forums for political action, defining professional commitments or orientations" (1995, p. 52). My observations at the deep ecology workshops affirm Thomashow's suspicion—ecological identities do inform decisions about work and about religious and social group affiliations.

Why would the the ecological identities of deep ecologists matter? Deep ecologists, at least those at the workshops, are rather unrepresentative of the broader society. The economically and racially homogeneous participants at the workshops were those, like myself, with the opportunities to pursue self-actualization and ecological harmony through nurturing ecological identities. Why not study the existence of ecological identities in other groups or in society as a whole?

The decision to study this unique group is justified for two reasons. First, in deep ecologists ecological identities achieve a prevalence and importance that most likely exceeds that in other groups. The average person's commitment to recycling may be a function of an ecological identity, but it is an identity that is infrequently drawn upon and kept isolated from other identities. In deep ecologists there is a greater integration of ecological identities throughout the self. This facilitates an understanding of how individuals work to create a coherent and meaningful self that incorporates and integrates ecological identities with other identities.

Second, recycling as an ecological identity is not difficult to sustain. Recycling has become so normalized in many communities that choosing not to recycle may threaten one's identity as a good neighbor or community member. Examining the self-concepts of deep ecologists forces us to consider the challenges of sustaining such identities, as discussed earlier. Deep ecologists better reveal the link between the self-concept and the social structure (House, 1981; Howard, 1991). An exploration of the constellation of identities that make up the self, especially when ecological identities are distributed throughout the self, reveals both the barriers to the development of ecological identities and the forces undermining existing ecological identities.

Given these barriers, one might argue that ecological identities are not "ecological" at all, but rather manifestations of social movement, political, or moral identities. In McAdam's (1988) study of the Freedom Summer campaign of 1964, the participants developed new activist identities through their political actions and then struggled to fit in with family and friends who could not understand these identities. Similarly, Lichterman (1996) has found that U.S. Green Party members have incorporated their sense of moral and social responsibility into their roles as political activists. In both of these cases activism seems to derive from the intersection of existing identities, value orientations, and life histories. Whether ecological identities are similarly derived, or emerge from experiences in nature, their potential for motivating behavior change is the same. In fact, since not everyone has access to nature and wilderness experiences that can forge ecological identities, the ability of this identity to emerge from the intersection of political, moral, or other identities is essential. In short, rather than concerning ourselves with the precise origin of ecological identities, we might do better to focus on the ways that current social structures and social meanings prevent ecological identities from becoming more important and more salient identities in a wider range of individuals.

Theoretical, Methodological, and Applied Implications

One intent of this research was to demonstrate that viewing environmental problems as microlevel problems of social interaction allows us to draw on social-psychological tools to explore how ecological identities are constructed, negotiated, and maintained. James (1890) and Mead (1934) provide strong theoretical bases not just for the notion of ecological identities, but also for the argument that such identities are at least in part social identities. By drawing on more recent developments in identity theory, I have attempted to argue that conceptualizing ecological identities as social identities opens up the possibility of exploring how ecological identities emerge out of, and are constrained and supported by, other social identities.

Methodologically, I have incorporated qualitative approaches that can tap the ways that individuals experience their relationship to nature, how this shapes their ecological identities, and how these identities are experienced relative to other identities. While the TST provided a relatively unstructured format for doing this, new methodological approaches are needed that integrate the TST's open-endedness with a more structured approach in order to systematically assess the relationship of ecological identities to other identities.

More pragmatically, the findings presented here may be of use to environmental educators. If we view environmental education as the process of helping individuals to develop ecological identities, then the activities of the deep ecology workshops are one example of the sorts of exercises that can develop and nurture ecological identities. Some of the more experiential forms of learning employed by deep ecologists, such as role playing, could be adapted to mainstream environmental education curricula. Employers might retain more of their ecologically minded employees by offering more opportunities for acting on ecological identities without feeling threatened. Such opportunities could range from company-sponsored nature outings to the formation of committees or decision-making procedures that specifically solicit ideas from ecological perspectives. Finally, religions also stand to benefit by creating more welcoming spaces for their members with ecological identities.

In each of these instances, we see once again how ecological identities are ultimately social identities. Environmental education, for example, relies on the social identities of teachers and students to convey the ecological meanings of social behaviors such as recycling. Ecological identities also are social identities in the sense that they must be nurtured and affirmed through interaction with social others. The implication is that in addition to developing ecological identities, environmental education also needs to give individuals the strategies, such as those employed by deep ecologists, for maintaining ecological identities.

Emphasizing the need to cope with social relationships that do not recognize ecological identities returns us to the notion that environmental problems are more accurately understood as problems of social organization. Current forms of social organization fail to make sense of our simultaneous existence "in irreducible realities: the obdurate physical world that is there and the symbolic institutions of social life" (Weigert, 1997, p. 47).

The self represents one of the ways in which we socially organize ourselves. Given this, ecological identities may represent the first step toward developing new forms of social organization. The significance of ecological identities will depend on our ability to create social institutions that not only accommodate ecological identities, but also recognize and affirm them. In the meantime, we need a better understanding of strategies for maintaining ecological identities, such as those employed by deep ecologists.

References

Axelrod, L. J. (1994). Balancing personal needs with environmental preservation: Identifying the values that guide decisions in ecological dilemmas. *Journal of Social Issues, 50*, 85–104.

Axelrod, L. J., & Lehman, D. R. (1993). Responding to environmental concerns: What factors guide individual action? *Journal of Environmental Psychology, 13*, 149–519.

Bell, M. M. (1998). *An invitation to environmental sociology*. Thousand Oaks, Calif.: Pine Forge.

Bragg, E. A. (1996). Towards ecological self: Deep ecology meets constructionist self-theory. *Journal of Environmental Psychology, 16*, 93–108.

Burke, P. J. (1991). Identity processes and social stress. *American Sociological Review, 56*, 836–849.

Devall, B. (1988). *Simple in means, rich in ends: Practicing deep ecology*. Salt Lake City, Utah: Peregrine Smith.

Devall, B. (1993). *Living richly in an age of limits*. Salt Lake City, Utah: Peregrine Smith.

Devall, B., & Sessions, G. (1985). *Deep ecology: Living as if nature mattered*. Salt Lake City, Utah: Peregrine Smith.

Dunlap, R. E., Grieneeks, J. K., & Rokeach, M. (1983). Human values and pro-environmental behavior. In W. D. Conn (Ed.), *Energy and material resources: Attitudes, values, and public policy* (pp. 145–168). Boulder, Col.: Westview Press.

Foote, N. N. (1951). Identification as the basis for a theory of motivation. *American Sociological Review 26*, 14–21.

Freese, L., & Burke, P. J. (1994). Persons, identities, and social interaction. *Advances in Group Processes, 11*, 1–24.

Gamba, R. J., & Oskamp, S. (1994). Factors influencing community residents' participation in commingled curbside recycling programs. *Environment and Behavior, 26*, 587–612.

Gardner, G. T., & Stern, P. C. (1996). *Environmental problems and human behavior*. Boston: Allyn and Bacon.

Gecas, V., & Burke, P. J. (1995). Self and identity. In K. S. Cook, G. A. Fine, & J. S. House (Eds.), *Sociological perspectives on social psychology* (pp. 41–67). Boston: Allyn and Bacon.

Gordon, C. (1968). Self-conceptions: Configurations of content. In C. Gordon & K. J. Gergen (Eds.), The self in social interactions (Vol. I, pp. 115–136). New York: Wiley.

Gutierrez Karp, D. (1996). Values and their effect on pro-environmental behavior. *Environment and Behavior, 28*, 111–133.

Hines, J. M., Hungerford, H. R., & Tomera, A. N. (1986). Analysis and synthesis of research on responsible environmental behavior: A meta-analysis. *Journal of Environmental Education, 18,* 1–8.

House, J. S. (1981). Social structure and personality. In M. Rosenberg & R. Turner (Eds.), *Social psychology: Sociological perspectives* (pp. 525–561). New York: Basic Books.

Howard, J. (1991). From changing selves toward changing society. In J. A. Howard & P. L. Callero (Eds.), *The self-society dynamic: Cognition, emotion, and action* (pp. 209–237). Cambridge: Cambridge University Press.

James, W. (1890). *The principles of psychology.* New York: Holt.

Kuhn, M. H., & McPartland, T. S. (1954). An empirical investigation of self attitudes. *American Sociological Review, 19,* 68–76.

Lichterman, P. (1996). *The search for political community: American activists reinventing commitment.* New York: Cambridge University Press.

McAdam, D. (1988). *Freedom summer.* New York: Oxford University Press.

McCall, G. J., & Simmons, J. L. (1966). *Identities and interactions* (Rev. ed.) New York: Free Press.

Mead, G. H. (1934). *Mind, self, and society.* Chicago: University of Chicago Press.

Naess, A. (1989). *Ecology, community and lifestyle* (D. Rothenberg, Trans. and Ed.) New York: Cambridge University Press.

Rosenberg, M. (1979). *Conceiving the self.* New York: Basic Books.

Schultz, P. W., & Oskamp, S. (1996). Effort as a moderator of the attitude-behavior relationship: General environmental concern and recycling. *Social Psychology Quarterly, 59,* 375–383.

Schultz, P. W., Oskamp, S., & Mainieri, T. (1995). Who recycles and when: A review of personal and situational factors. *Journal of Environmental Psychology, 15,* 105–121.

Seed, J., Macy, J., Fleming, P., & Naess, A. (1988). *Thinking like a mountain: Towards a council of all beings.* Philadelphia: New Society Publishers.

Statham, A. (1995). Environmental identity: Symbols in cultural change. *Studies in Symbolic Interaction, 17,* 207–240.

Stern, P. C., & Dietz, T. (1994). The value basis of environmental concern. *Journal of Social Issues, 50,* 65–84.

Stern, P. C., Dietz, T., Kalof, L., & Guagnano, G. A. (1995). Values, beliefs, and proenvironmental action: Attitude formation toward emergent attitude objects. *Journal of Applied Social Psychology, 25,* 1611–1623.

Stone, G. P. (1962). Appearance and the self. In A. M. Rose (Ed.), *Human behavior and social processes.* Boston: Houghton Mifflin.

Stryker, S. (1980). *Symbolic interactionism: A social structural version.* Menlo Park, Calif.: Benjamin/Cummings.

Stryker, S. (1991). Exploring the relevance of social cognition for the relationship of self and society: Linking the cognitive perspective and identity theory. In J. A. Howard & P. L. Callero (Eds.), *The self-society dynamic: Cognition, emotion, and action* (pp. 19–41). Cambridge: Cambridge University Press.

Thomashow, M. (1995). *Ecological identity: Becoming a reflective environmentalist*. Cambridge, Mass.: MIT Press.

Weigert, A. J. (1991). Transverse interaction: A pragmatic perspective on environment as other. *Symbolic Interaction, 14,* 353–363.

Weigert, A. J. (1997). *Self, interaction, and natural environment: Refocusing our eyesight.* Albany: State University of New York Press.

Zimmerman, M. E. (1993). Rethinking the Heidegger-deep ecology relationship. *Environmental Ethics, 15,* 195–224.

15

Identity and Sustained Environmental Practice

Willett Kempton and Dorothy C. Holland

Surveys and interviews since the 1970s have shown that many or most persons in the general population know about environmental problems, profess concern about them, and agree with the goals of the environmental movement (e.g., Dunlap, 1992, 2000; Brechin & Kempton, 1994; Kempton, Boster, & Hartley, 1996). So why do so few carry these concerns into sustained environmental practice? This question has been the subject of academic research as well as speculation by interested lay observers. One frequently proffered answer is that talk is cheap—people will say what seems socially desirable, but will not really expend effort to act. A second frequent answer is that although people have environmental values, other values such as having a sporty car or displaying one's social status are more important. The question posed, as well as both answers, make the theoretical supposition that values and beliefs are sufficient to influence behavior. Our perspective is that an equally or greater causal factor is the individual's identity, which leads to behavior consistent with that identity. Thus we seek to understand the ways in which a person develops a sense of him- or herself as an environmental actor, and how such an identity, once formed, affects behavioral practice.

Given our goal of understanding the causes of environmental practice, we reasoned that our empirical research should not sample the population at random since very active people are infrequent and sparsely distributed. Rather, we have weighted our sample toward members of local environmental groups, who are known to be taking action, with small comparison samples drawn from the general public and members of national environmental groups. We asked these informants to describe the environmental practices they followed, their self-ascribed identity, their recollections of events that affected their awareness of environmental problems, and their life history of involvement in the environmental movement.

The identities yielded by this process were "social environmental identities," that is, self-definitions with respect to one's reference group, the environmental movement, the government, the marketplace, and lifestyle choice. Social environmental identities locate a person as an environmentalist, or a particular type of environmentalist, in a context of persons, groups, and struggles. One's social environmental identity also contrasts with that of others—others who have different identities, take different roles within or opposed to the movement, and carry out different actions. Our social identity orientation did not, therefore, focus on identities that relate a person to the natural world, sometimes called "ecological identity," unless the informants brought such identities up in our free listing questions.

We find that environmental action, as defined by people themselves, is multisided. We distinguish two major areas of action—civic action and cultural reform—and describe here two types within each area of action. From the point of view of the first—civic action—corporate behavior and weak or inactive government are seen as the major problems, and an environmentalist works with the civic sector, attempting to move government or sometimes corporations. Within the civic sector there are a multitude of identities and types of action, of which we will describe two, for example: (1) civic action "at a distance" as member of a large "mail-in" environmental group, whose activities include reading a newsletter, sending membership dues, etc., versus (2) civic action as a participant in a local group, organizing, meeting face to face, and petitioning local government.

The second major area of action responds to threats perceived to come from within the individual, community, and consumer culture rather than from external sources such as corporations. People with this latter perspective claim to be environmentalists by virtue of their attempts to make their own lives examples of good environmental practice, for instance, by recycling containers and properly disposing of motor oil. For the individual self-reformer, such behaviors are central, and sufficient, to claim one's identity as an environmentalist. For the social reformer, action includes efforts to convince others to change their practices as well. Whether working at the individual or social level, we refer to persons with this second type of environmental identity as cultural reformers. Cultural reformers are producing new life-styles and, in their most organized forms, new communities.

In this chapter we review the literature on identity in the environmental movement and the causes of environmental action. We then draw contrasting cases from our interviews with a sample weighted toward members of environmental groups. The interviews cover identity, reported environmental actions, and a narrative of the informant's life history and development in the environmental movement. These data suggest a relationship between identity and action, illustrate the different forms of environmental action, and (via the narratives) illustrate the context and development of identity and action. Next we provide a theoretical perspective on identity development, which we think frames the qualitative data. Finally, we use codings of our interviews to measure the amount of variance in environmental action that can be explained by our identity measures.

Literature on Types of Activists

Many of the studies attempting to understand why only some people take environmental action are attribute-correlational. In such studies, first, a number of personal attributes are hypothesized that might lead individuals to take action—for example, environmental values, attitudes toward the environment, or various sociodemographic characteristics. Second, a survey is conducted asking about these attributes for the individual survey respondent, plus asking about their environmental action. Although our approach is more oriented to the processes that convert a nonactive person into an active one, these attribute-correlational studies do offer helpful background insights.

One of the best examples of a recent attribute-correlational study is that by Stern, Dietz, Abel, Guagano, and Kalof (1999). They examined six theories of causes of environmental behavior and found that this reported behavior was best explained by a combination of these theories, which included underlying values (i.e., altruism, unity with nature, family security), environmental beliefs and awareness of consequences, ascription of responsibility, and personal environmental norms ("I feel a personal obligation to . . . prevent climate change."). Like us, they divided environmental behavior into several types. Stern et al. distinguish "citizenship actions" such as writing a letter to Congress, "personal sphere behavior" such as recycling, and "policy acceptance," such as expressing willingness to have regulations that would impose on the respondent. (We similarly distinguish citizenship actions from personal behavior; on the other hand, we feel that policy acceptance is closer to an attitude than a behavior.)

In describing new social movements such as the environmental and women's movements, Melucci (1988) and others contrast them with older social movements, such as the labor movement. Whereas the older movements focused on gaining political power as conventionally defined, environmentalism and the other new movements are engaged in cultural politics as well. In these movements, cultural production is just as significant as political activity (Melucci, 1988).

To regard activists only in relation to the attempts of environmentalists to affect government policies, laws, and programs is to miss a good bit of their force. These movements also could produce systemic change through their social and cultural dimensions. They produce and experiment with new "cultural codes" that "constitute the submerged activity of the contemporary movement networks and the condition for their visible action" (Melucci, 1988, p. 337). (We correspondingly find in our data a second type of action that we call "cultural reform." Cultural reform is directed at changing one's own life-style and private environmental practices, and possibly that of others.)

Flacks (1988) presents a typology of political activism with three categories: ordinary citizens, citizen activists, and elites. The first are what Flacks calls "ordinary people," the majority of the population, who choose to remain in the private realm of everyday life. They will act politically only if they perceive that their daily life is threatened, but will return to private life when that threat has been defeated. Citizen activists, in Flacks's typology, have arranged their private lives so that they can put a larger amount of energy into political activity. They are predisposed to action owing to high education and a middle- or upper-class background. Flacks contrasts them with elites, for whom activism is part of their jobs. He calls them "activists" in the sense that they are government or business leaders who "make history" as part of their jobs.

Following Flacks's lead, but making one of the few empirical studies of pathways to environmental activism, Aronson (1993) identified a number of core members of citizen groups working to change government policy and corporate practices. Aronson's activists were involved in the grassroots hazardous waste movement. As with other researchers focusing on components of the movement that began to develop in the late 1970s and 1980s (e.g., environmental justice and the toxics movement), and counter to Flacks's statement, Aronson did not find a predisposition to activism among those of higher class and education levels. He describes the people he interviewed as predominately "lower-middle class and working class." Moreover, consistent with the social practice theory that informs our work (Holland, Lachiotte, Skinner, & Cain, 1998), the process of becoming an activist is almost the opposite of having a predisposition. Aronson found that "political activity *precedes* an activist political consciousness and it *initiates* the transformation of an individual into an activist" (Aronson, 1993, p. 64, emphasis in original). Citing the Marxist concept of "praxis," Aronson theorizes that "*action changes consciousness* of the self and the other objects in the situation, and this in turn affects future actions" (1993, p. 76, emphasis in original). (As we will show, our data substantiate Aronson's view of activity preceding a consciousness as an activist. Also, we focus more on identity as a mediating variable interacting with both action and what Aronson calls "political consciousness.")

Research Methods

The research reported here consisted of three major parts. First, we made a comprehensive inventory of local environmental groups in two areas in the United States. We found 132 local environmental groups on the Delmarva Peninsula and 434 in the state of North Carolina, which is more than ten times the number listed in a national directory of environmental groups (Kempton, Holland, Bunting-Howarth, Hannan, & Payne, 2001). That comprehensive inventory found that the most common type of local environmental group was what we called "civic groups," who were politically involved in their communities and engaged with local government.

For intensive study, we selected a judgment sample that included a variety of the civic groups shown by our inventory to be most prevalent, but also included at least one each of most of the major types identified in the inventory. The following list summarizes the ten types of environmental groups, plus the two comparison groups (environmental scientists and members of the public).

Radical: local; direct action, confrontational; biocentric, sometimes anarchist ideology

Civic: local (sometimes organized around one community or environmental problem); political action and networking; very diverse issues

National: national; mail-in membership, advocacy by staff

Life-style: local; focus on improving members' sustainable living practices and consumer actions

Environmental justice: groups that oppose environmental threats to the quality of life for racial minorities or poor people

Students: local; high school and college environmental clubs

Conservationists: local; natural resource user groups; focus on land conservation and habitat protection

Wise use: local; resource users; focus on maintaining human use rights

Resource user group: local and occupational; for our sample, these were fisheries groups; work to preserve stock but with minimal regulations; equate healthy ecosystem with maintaining healthy fish harvests

Scientists: EPA environmental professionals; science training (control group)

Public: sample of local population of adults drawn from phone books (control group)

Our sample, 20 of the 566 groups inventoried, is described in Kitchell, Kempton, Holland, and Tesch (2000), which explains our grouping in more detail and lists the specific 20 local group samples. Note that, unlike Brulle (1996) and related work, we distinguish environmental groups not only on the basis of their discourses but also their members' actions and identities—resulting in a rather different clustering of groups. For example, we cluster all national groups into one type because regardless of the groups' discourses, most members take very limited action (they read newsletters and write checks) and members' identities seem little different regardless of which group they belong to. On the other hand, we distinguish among conservationists, resource users, and wise use groups because, despite similarities in their discourses, the members' identities and actions are very different.

In each local group selected for more intensive study, approximately eight of the most active members were interviewed regarding their history of identification and involvement with the environmental movement and environmental action. These identity interviews yielded complete information on 159 members.

The national environmental group sample was drawn from a list of members provided to us by a national environmental organization. The public was taken from a sample randomly drawn from phone books in a western North Carolina county and the northern half of the Delmarva Peninsula. The environmental scientists were an opportunity sample from a US EPA office in North Carolina.

Examples Relating Identity and Environmental Action

To illustrate variation in our data on identity and action, table 15.1 presents five examples of individuals selected from our corpus of 159 indepth interviews. The first column lists each individual's pseudonym, the type of group they belong to, and brief demographics. The second column lists the identities given by the informant, in the order they gave them, in response to the so-called Twenty Statements test. This test, introduced a half-century ago by Kuhn and MacPartland (1954), asks for up to twenty responses to the question "Who am I?". It has become a standard instrument for eliciting identity. The rightmost column lists the environmental actions individuals said they took, numbered to indicate their ranking ordered by which actions they felt had the most effect. In short, the table compares the type of group they belong to with their identities and with their reported actions. One member of the public is shown (pseudonym "Bruce") followed by members of four contrasting environmental groups.

Table 15.1 illustrates the consistency we find in many cases. Specifically, we see consistency among the type of group a person belongs to, their self-expressed identity, and the types of environmental actions they report performing. The selected cases in table 15.1 are not intended to prove that relationship—we leave that to our statistical analysis later in this chapter. Table 15.1 is intended to illustrate with qualitative, textual data the types of correspondences we find, as described later.

Table 15.1
Identity and environmental action elicited from five interviews

Pseudonym, group type, demographics	Identities	Environmental actions reported
Bruce (public sample), white male, 50, 3 years of college; mainframe computer operator	American Logical Veteran Father Fair Demanding	1 Recycle oil, plastic, and aluminum (listed separately) 2 Discontinued use of two-stroke outboard motors 3 Catch and release fish (sport fishing)
Jerry (national group), white male, 65, master's degree, financial planner	Retired school teacher Financial planner Father Grandfather Concerned individual regarding people and the planet Love music Enjoy some sports Walking Enjoy gardening Enjoy traveling I feel I'm a religious person in that I care about people, society, getting along, etc. Family is very important	1 Support groups, such as the local nature center 2 Recycle 3 Won't deliberately purchase things, such as a sport-utility vehicle, or commit detrimental acts

Naomi (cultural reform group for consumption reductions), black female, 34, J.D, city asst. solicitor

Woman
Spiritual being
Black person
Writer
Lawyer
Daughter
Sister
Lover
Swimmer
Student
Artist
Thinker
Volunteer
Gardener
Friend
Listener
Consumer
Commuter
Cat person
Runner

1 Use few chemicals such as insecticides and fertilizer
2 Recycle
3 Not eat meat
4 Buy less
5 Create less trash
6 Compost

Table 15.1 (continued)

Pseudonym, group type, demographics	Identities	Environmental actions reported
Linda (local civic group), white female, 50's, some college, part-time businessperson	Wife Grandmother Daughter Sister Friend Businesswoman Environmentalist Politician Volunteer Writer/poet Gardener Honest Loyal Passionate Fighter Member of my church Shopper Sensitive Angry	1 Involved in grassroots movements 2 Attend as many diverse meetings as possible 3 Write to elected officials to bring issues to their attention 4 Stay in contact with EPA, MDE [Maryland Department of the Environment], and other state or federal departments 5 Taking plant life for granted 6 Recycle more 7 Read any articles on environmental issues 8 Watch movies and programs to learn resources to pass on

	Identity	Practices
Garth (local civic group doing demonstrations), white male, 47, B.A., consultant for business	Activist Consultant Writer Friend of a lot of people Do-it-yourself person Researcher or analyst	1 Advocate protection of environment 2 Use less energy 3 Recycle wastes 4 Don't buy lots of new stuff 5 Try to use safe, nontoxic products 6 Minimize meaningless consumption 7 Vote against Republicans
Jim (radical group), white male, 41, 6 years of college, consultant for environmental campaigns	An environmentalist A revolutionary An activist Direct actionist Radical An EarthFirster Anarchist Revolutionary ecologist Anticapitalist Enemy of the state Hell-raiser Hippie, pinko, commie scum A human Part of real counter-culture Nature lover Tree hugger Environmental wacko (proud of it!) A watermelon (green on outside, red and black on inside)	1 Educate other people about environmental issues (workshops, talks, leaflets) 2 Engage in nonviolent direct actions and protests 3 Consume less 4 Eat less meat (I try not to buy it) 5 Weak attempts at healthy eating and diet 6 Reuse stuff 7 Recycle everything 8 Carpool when traveling 9 Ride a bicycle

Bruce, from the public (nongroup members), is a Vietnam veteran who is suspicious of the government. He does not list any identities related to the environmental movement. His reported environmental actions are recycling plus two actions he performs as a sports fisherman. (He began recycling oil after he noticed that nothing grew on the spot in his back yard where he was dumping it.)

Jerry, the member of a national environmental group, describes himself as "concerned individual regarding people and the planet." His first action is to "support groups," that is, by donations and his membership; the other two are consumer actions. The interview makes it clear that he is environmentally concerned, and his mail-in group membership and donations are actions that are very much that of a concerned individual, rather than someone acting in concert with, or in opposition to, others in his community.

Naomi, the member of the consumption-reduction group, like the public example, gives no environmental identities; her environmental actions consist primarily of controlling her own consumption, although she also writes Congress and "advocates," perhaps in personal conversations.

Linda, the civic group member, although leading with kinship terms, identifies herself as an "environmentalist" and gives participatory civic identities such as politician and volunteer. Her four highest-ranked environmental actions all involve local political activities, with only one action based on personal consumption (recycling).

Garth, the organizer of a civic group that plans demonstrations, leads his identity list with "activist." His list of actions begins with "advocate protection of the environment," followed by several personal consumption items.

Jim, the radical group member, when asked to identify who he was, gave virtually all movement terms. Although terms such as "commie scum" and "environmental wacko" suggest that some answers are tongue-in-cheek, it is striking that when asked "Who am I?," his first term was "environmentalist," and all the terms are related to his movement identities. Kinship terms do not even appear, as is also the case for the activist Garth but unlike the responses of most informants, including all the other informants listed in table 15.1. Jim's actions combine political or protest actions as well as personal consumption ones, with the consumption actions (like the political ones) being more dramatic than those of any other informant (e.g., "recycle everything" rather than "recycle").

In sum, by examining identities and actions across individuals in different environmental groups, we see the variation in answers and the types of correspondence possible among identity, action, and group membership. In the next section we look in more detail at one of these cases in order to contextualize these data and obtain some clues as to the underlying processes.

Case Study of an Environmentalist

This section describes one of our informants, Garth, through his narrative history of environmental involvement. This case is useful in that he remembers and articulates key aspects of his development as a particular type of environmentalist. In particular, he recalls his early action that led him to define himself as an environmentalist, and his later actions within the environmental movement that led him to create a new group, which he then defined as a "real environmental group" in contrast to other groups. Although the space required for case studies allows us to include only one in this volume, cases of cultural reformers, national group members, or natural resource group members would illustrate differing identities and paths to action; their contrasts with Garth can be at least surmised by comparing the informants in table 15.1.

Garth, Development of a Local Activist As someone who has contributed many of the 47 years of his life to environmentalism, Garth qualifies for the label of "career activist." He works as a consultant, but he is heavily occupied with his activism. He is an organizer of an umbrella group, "Green Delaware," which, more than any other local group, has carried out demonstrations and other more confrontational tactics in the state of Delaware.

Garth went through a series of steps before becoming an activist-organizer. Appropriately to his currently confrontational style, his first recollection of environmental awareness was an argument with his father over whether his father's firm should be trying to evade environmental regulations. As an adult, the first time he started thinking of himself as a strong environmentalist was, he says, "when I was consulting for

Du Pont and I started to . . . take the risks of arguing . . . [for environmental protection] . . . that was probably when I started to have a self-identity as an environmentalist." Up until this point, he worked from within his job to try to convince companies to do more for the environment. It is a case of action preceding identity; that is, he suggests that his "taking the risks of arguing" led to an identity as an environmentalist.

His entry into organizing and participating in a significant citizen's group came serendipitously. As he recalls, he was thrust into an organizing position when he was asked to be on the executive committee of the Delaware Sierra Club. He says that they "just wanted a warm body" for the committee, but "I started to see . . . 'Hey, why don't we try to do something with this group, why don't we try to be a presence?' And that was the catalyst really, for whatever activist work I've done." Again, engagement in the practice of environmentalism seemed to advance his strategy and his view of possible actions.

The Sierra Club eventually asked Garth to resign. In another interview, he describes the formation and purpose of his own group, Green Delaware. He and three other people, each of whom initiated some actions under separate organizations, connected:

Well we had [been] working together on a variety of issues and . . . it seemed . . . that we could achieve better public recognition if we had a single name for the activity . . . we are all people who'd believe in, who are . . . willing to picket, willing to hold press conferences, willing to try and kick up a fuss. Which is not characteristic of, for example, in Delaware, the Audubon or the Sierra Club. They just don't do that. So we all had kind of a common, a common mode of operation, if you like.

Garth now views all groups in the area except his own as "not real environmental groups," saying "everybody in the state, including the environmental organizations, are in their pocket [of corporations]." As he started to organize his own group, he recalls a crucial step as meeting three activists in Delaware. Of one in particular, Jimmy (a pseudonym), he says: "I was influenced by him. . . . Jimmy was the only one who was really willing to share his knowledge and was encouraging. Arthur's (pseudonym) interest in water, Jimmy in air, me in energy . . . so we tried to kind of pull that all together as Green Delaware."

The formation of their own environmental group, Green Delaware, corresponds to a refinement of their environmental identity. Note that

Garth distinguishes it as a "real environmental group," and by extension, presumably they see themselves as "real environmentlists" in contrast to others in the environmental movement.

Garth's identity terms begin with "activist." His action list starts with advocacy, as he thought it had the most effect, but also included many personal actions. These can be seen in table 15.1. This list contrasts with his narrative of involvement in the movement, in which he focused almost exclusively on political and organizational activities.

Although he is more individualistic and confrontational than most leaders of civic action groups, Garth offers an example of having gone through changes in his own identity and his strategy for action, affected by his membership in groups and ultimately creating his own local environmental group.

Three General Aspects of Identity Development
We now introduce a theoretical perspective on identity formation to frame and explain some of the rich data in table 15.1 and the case study. Holland et al. (1998) in their book, *Identity and Agency in Cultural Worlds*, describe several general interrelated aspects of identity formation. As people develop a sense of themselves as actors in a cultural world, three changes occur. One is that the cultural world—in this case, the world of environmental action—becomes much more salient. Apprehension of environmental threat and action becomes more acute and ever present.

A second change is identification of oneself as an actor in the world of environmental threat and action. One has to learn enough to imagine taking part in the world, to care about the consequences of one's actions, and to care about evaluations of oneself in relation to the world. Identification involves investing one's self, taking some responsibility, being answerable to or becoming open to being called to account for one's actions.

The third change is practical knowledge—knowledge obtained through action. As a person engages in environmental actions, he or she gains familiarity with the social relations and practical activities of environmental action. Those who have progressed in their practical knowledge reach the point of being able to produce guides for others and to mentor neophytes as they were once mentored.

In summary, these three aspects of identity development are salience, identity, and knowledge from doing. We describe each of these, with examples. Holland et al. (1998) argue that these dimensions occur during identity formation. Our life history data suggest additional aspects of the identity processes. However, these data show that the processes do not necessarily occur in the order given. Furthermore, there are important reformulations specific to participation in the world of environmental action, and we identify barriers that can prevent sustained environmental practice. That is, when we examine data from this movement in particular, we find specific, important reformulations and barriers to those reformulations, which could also be considered to be cases of arrested identity development.

Salience   We found many instances where our interviewees described environmental problems becoming more apparent to them. They often use the word "aware" or "waking up" to describe one of the more important parts of this reformulation. They became "aware."

Often the triggering event was a direct experience with environmental destruction in their neighborhood and the threat of displacement from a familiar place or way of life. For example, one woman indicated her rising apprehension that her valley was becoming unlivable:

This valley was absolutely so beautiful when we first moved out here. It was absolutely beautiful. Then, the industry started moving in to the valley, and I can deal with most of the industry until it gets to the point that it is a tremendously polluting, a heavy polluting industry. And, I think the asphalt plant [that her group had been opposing] definitely fits that definition. So, I guess that's the point at which we stood up and said "Enough is enough!" At least you have to be able to breathe. You may not be able to look out and see a beautiful landscape anymore, but you at least have to be able to breathe!—Andrea, Citizens Unite

From there her initial awareness of a single instance of local environmental damage became linked to a larger (conceptual) picture of the pervasiveness of pollution. It is not just the valley that is becoming unlivable.

I have a daughter with asthma, and as I read more and more about asthma, I realize that it's on the rise. And that's the big slogan of the American Lung Association, "Asthma is on the Rise." Well, you wonder why with all the technology and medical advancements, why a disease like this would be on the rise. Well, the more reading you do, the more you find out that there are environmental issues involved. According to the American Lung Association, there is an increase in asthma . . . when the ozone level is up. So, it kind of points to air pollution and problems with quality of the air.—Andrea, Citizens Unite

For others, the transition to awareness was not caused by a pending threat or dislocation, but rather an experience or event that interrupted the taken for grantedness of the physical world around them. Naomi, mentioned earlier, told about her experience of gardening. Similar to Naomi and consistent with the widely held belief that experiencing nature creates environmentalists, a woman in an Earth First! Group in North Carolina said:

I learned a lot about nature. I think that was a lot of what motivated me to care about environmental destruction—just learning plants, their medicinal properties, learning that I'm connected with this planet, learning that that thing is ginger and I can make tea out of it, and it's growing right next to me. And that was a lot of the Earth First! gave me, because there were a lot of nature walks, and herbal walks, and things like that.—Stella, Ruckus Society

Identification with the World of Environmental Action Identification means experiencing oneself as an actor in the world of environmental action. It involves accepting responsibility or answerability and caring about how one's (group's) behavior fits what one claims to be. For many of those becoming oriented to citizen activism and the environmental practices that entails, a first reformulation is to realize that government may not do the job, or that corporations are not necessarily benign; this is often experienced as disillusionment. A subsequent step is what some called "empowerment"—acquiring a sense of agency, whether alone or as a group member.

In some cases, the person seems to have taken on a role enabling, but not requiring, environmental action. They take on the role, act accordingly, are identified by others as an environmentalist, and thus begin to think of themselves as environmentalists. This is the opposite of values or beliefs leading to action; rather, it is a case of action reinforcing or expanding environmental beliefs. One example is Garth, cited earlier, who was put on the Sierra Club executive committee because they were looking for a "warm body," but who then saw it as an opportunity for action.

One example of actions preceding identity is taken from our public sample. This person, Eunice, performs environmental actions, such as recycling, because they are habits formed from living in India. From an early age, she was taught to save paper and bags until they were no longer usable. Similarly, most travel in India was by public transportation, or bicycles, or on foot. Eunice did not think of herself as an environmentalist until she moved to Canada and then the United States and learned the connotations ascribed to her habits—saving paper and recycling cans and bottles was considered good for the Earth.

Another example comes from an earlier stage in life, from Julia, a member of a high school environmental group. To the question, "When did you first start thinking about yourself as an environmentalist?," she replied:

I guess after I joined Nature Society. Before I was an environmentalist in talk but not in action. You know, I would argue about it and talk about things like that, but I wouldn't do anything. So I guess I was an environmentalist in that sense, 'cuz I was aware of it, but I wasn't doing anything to change it. So I guess now I'm working, you know, in a small way to make a difference. So I guess that's an official environmentalist.—Julia, Newark High School Nature Society

Furthermore, once actions have caused a reformulation of one's self as an environmentalist, that in turn affects one's perceptions of the world of environmental damage:

This campaign [against an asphalt plant] has led me into being a fairly strong environmentalist. I see things now that I wouldn't have ordinarily seen in the past. When I see things happening—big exhaust coming out of trucks, the first thing I think of now is how much sulfur dioxide is going into the atmosphere, whereas before it was just smoke coming out of an exhaust.—Ray, Citizens Unite

And it becomes central to self-definition and the meaning of one's life:

If I couldn't work on environmental issues, if I couldn't be involved in the community, if I couldn't be involved in making changes and differences, there's no reason for me to be. This is part now that my children are grown, uh, that I'm retired, that I have the time to get involved in the community, it actually is my vocation at this point. It's something that I do. It's something that I identify with very strongly, and that would be a tremendous loss to me.—Alexis, Haztrak

Also, group participation can enhance the sense of empowerment:

There was this group of wonderful people that I loved so much that were moving more into an environmental direction, and a lot of them were sort of waking up. This woman Julienne, a really, really awesome, an artist, who did silk-screened stuff, she was just really starting to get active. She had sort of been like a brain-dead hippie type for a long time. And all of these people were all just coming to this awareness that at least we were going to do something about it.—Stella, Ruckus Society

Increasing Practical Knowledge and Other Resources for Action   As environmentalism becomes more salient through practice and a person identifies her- or himself more in relation to environmental action, she or he becomes more knowledgeable about the environmental movement and how to carry out actions and engage in environmental practice. For example, in an earlier quotation, Ray reported that he became much more knowledgeable about air quality. He also learned which levels of government were important to lobby on air quality issues.

Practical knowledge may be acquired from a group or from an individual mentor. In some cases, like that of Garth, there is a specific person who is identified as sharing their knowledge. Knowledge may be acquired through group activities and hearing stories from more experienced group members. In the case of Citizens Unite, experienced members of other environmental groups specifically attended Citizens Unite meetings and tutored them on the tasks that they would have to accomplish and how to organize their groups for those tasks.

Samantha provides another example in her description of her summer camping with other Earth First! members as a peer-to-peer learning experience:

A lot of it was education that I was getting from the other environmentalists that I was camped with. About lead mining, and sometimes we would drive by these huge plants and factories and stuff. But, what was really empowering in the summer, was just a group of kids, all around my age, between 19 and 23, or something, maybe 20 of US. . . . It was just a really great effort to learn a lot, and people did learn a lot about organizing that summer. And all this other stuff that we were new to, we were all inexperienced, most of us were. It was really empowering.—Samantha, Ruckus/Earth First!

Other examples of how individuals acquire practical knowledge, and other structures to facilitate environmental action, can be found in other publications from this project (Kitchell, Hannan, & Kempton, 2000).

Other Changes, Barriers, and Reformulations   While the three dimensions of identity formation discussed here are relevant to the development of environmentalists, they do not specify the aspects of the process that come about because the identity is developing in the context of this particular social movement. The life histories of environmentalists reveal a culturally and socially imposed inertia of nonenvironmental identity and practice. We refer to specific beliefs or identities causing this inertia as "barriers."

The development of an environmental identity and a sustained environmental practice involves moving past this inertia, often bit by bit with the help of others. It involves taking multiple steps into the world of environmental threat and action. We refer to these changes in identity, concepts of action, etc. as "reformulations." The reformulations differ, depending upon the kind of group and the kind of environmental practice. They also can differ by region (the threat of displacement moves rural people quickly through the salience transition), and by ethnicity (black environmental justice activists may not need to make a reformulation common among whites—questioning whether the government is looking our for their interests). Also, for those in our research who were involved in citizen activism, the reformulations involved were by and large different from those involved in culture or self-reform. An initial analysis of these barriers and reformulations can be found in Kitchell, Kempton, Holland, and Tesch (2000); more will appear in a manuscript now being prepared by Kempton and Holland.

Correlation of Identity with Environmental Action

The example cases and quotations suggest a connection between identity and environmental action. Our theoretical perspective fits many of these reports into an overall pattern that leads to environmental action of various types. However, the data given earlier are from selected cases. For a more systematic examination, we used the coding of all 159 informants.

Our coding categorized identities into major types. For example, a "consumer" identity was coded when the informant gave terms such as *shopper, bargain hunter,* or *conscientious consumer.*

To code actions, we categorized environmental actions into citizen actions (write officials, attend meetings, etc.) versus consumer actions (recycling, ride a bicycle, don't buy harmful products, etc.). We also counted the number of groups they belong to as a variable that could be considered a cause (group membership leads to action) or a result (environmental identity leads to group membership). As a contrast to the identity terms related to the environmental movement, community, or consumption, we also counted kin terms, which appeared frequently in the identity data but which we did not expect to be correlated with environmental action. Finally, the identity terms *gardener* and *animal lover* (sometimes given as *animal/plant lover*) were not anticipated, but seemed to occur frequently in the identity terms of environmentalists, so we included them in the correlation test. The coding is described in more detail in Kitchell, Kempton, Holland, and Tesch (2000), who use these data to distinguish among environmental groups (but not to correlate identity with action, as we do here).

Table 15.2 shows that for the consumer actions, none of the identities correlate significantly with actions—even identifying oneself as an environmentalist has only a 0.13 correlation ($r = 0.13$, $p = .11$, n.s.) with consumer environmental actions.

Civic environmental actions, by contrast, are correlated significantly with the identities of "activist" ($r = 0.41$, $p < .0001$) and "environmentalist" ($r = 0.34$, $p < .0001$) and also with belonging to more environmental groups ($r = 0.34$, $p < .0001$) and with identifying oneself as an

Table 15.2
Correlation (Pearson's *r*) of environmental identities with reported environmental actions and group membership ($N = 159$ individuals interviewed)

	Environmental actions		
Identity	Consumer actions	Civic actions	# groups
Kin terms	−.01	.10	−.07
Consumer	−.01	.08	.09
Citizen	.08	.05	.04
Environmentalist	.13	.34***	.44***
Ecosystem role	.03	.15	.13
Activist	−.05	.41***	.15
Gardener	.03	.06	.25*
Animal lover	.14	.25*	.25**
Number of groups	.09	.34***	1.00

† $p < .05$, * $p < .01$, ** $p < .001$, *** $p < .0001$.

animal lover ($r = 0.25$, $p < .05$). Considering group memberships, we find the identities of environmentalist, gardener, and animal lover to be significantly correlated with the number of groups (see table 15.2).

On the other hand, identities related to kinship, being a consumer, or being a citizen have little or no relationship to environmental actions, whether consumer actions, civic actions, or joining more environmental groups. Although we did not expect kin terms to correlate with environmental action, one might plausibly expect people who label themselves with consumer identities (shopper, bargain hunter) to be more likely to take consumer actions, and those labeling themselves with citizen identities (voter, etc.) to be more likely to take citizenship actions. Table 15.2 shows that this was not the case.

Next we performed multiple regression with the dependent variable as civic actions and the independent variables as the eight identities given (not the number of groups). The overall multiple regression explained 27 percent of the variance ($F = 6.71$, $p < .001$). Only three identities contributed significantly. Again, they were activist ($B = 0.41$, $t = 4.24$, $p < .0001$), animal lover ($B = 0.45$, $t = 3.11$, $p < .005$), and environmentalist ($B = 0.15$, $t = 2.26$, $p < .05$). None of the other identities were significant at even the 0.2 level.

In a multiple regression on consumer actions as the dependent variable, the whole set of eight identities could explain only 6 percent of the variance ($F = 1.13$, n.s.). Nor did any single identity contribute significantly to the equation (even at the $p < .1$ level). It would be plausible that some of the variables predicting civic actions, such as activist or ecosystem role, would lead one to perform more environmental actions as a consumer. This does not appear to be the case.

Our negative finding, the lack of correlates of consumer actions, should not be considered the final word. The question about environmental action was open ended and persons engaged in more civic actions may have listed them first, leading to their truncating the consumer list. Also, the civic actions list is a more difficult test, in that everyone can recite consumer actions whether or not they do them, but the civic actions may not be known, or at least not as easily recalled, by individuals not doing them.

In interpreting our data, one should also recall that our sample consisted primarily of members of local environmental groups. Correlations that would appear when the samples were taken primarily from relatively uninvolved members of the public, or when equal numbers of the public and group members were compared, may not appear in these results. Thus it may be, for example, that the identities of activist, animal lover, and environmentlist together explain 27 percent of the variance in civic actions *only among environmental group members*. We are currently (in mid-2002) analyzing a survey with fixed questions on consumer and civic actions, using respondents that include the general public as well as environmental group members, which may shed more light on these questions.

In short, several types of identity seem to be correlated with civic actions for the environment. Although some of our case study data suggest relationships between identities and consumer actions, those relationships cannot be detected in our quantitative analysis of the interview data.

Conclusion

In our study of 159 individuals, mostly sampled from the membership of local environmental groups, we found a relationship between environmental identities and environmental actions taken. This can be seen in the qualitative data in that the identity terms the respondents listed are consistent with the environmental actions they listed. The relationship can also be seen in the narratives they used to describe themselves and the reformulations of their identities, as intertwined with actions taken and group involvement. The relationship between identity and action is confirmed in the quantitative data. For example, knowing only three identities—whether a person considers themself to be an activist, an environmentalist, and an animal lover—explains 27 percent of the variance in the number of civic environmental actions they take.

The truism that correlation does not prove causality seems especially to apply in interpreting the correlational data. Although one might traditionally think of identity as causing action, with group membership plausibly causing both identity and action, that linear model is not supported by the narratives and quotations we have presented. The narratives suggest that all three are mutually causal, more akin to positive feedback than to strict cause and effect.

Acknowledgments

We are grateful to many interviewers and field workers who collected the data for this project. The coding of the 159 identity interviews was done by Danielle Tesch and Anne Kitchell. Doug Christel assisted with analysis and manuscript preparation. We are also grateful to the many members of local environmental groups who gave their time to participate in interviews. This research is sponsored by grants SBR-9602016 and SBR-9615505 from the National Science Foundation, for which Kempton and Holland are the principal investigators.

References

Aronson, H. (1993). Becoming an environmental activist: The process of transformation from everyday life into making history in the hazardous waste movement. *Journal of Political and Military Sociology, 21*(Summer), 63–80.

Brechin, S. R., & Kempton, W. (1994). Global environmentalism: A challenge to the postmaterialism thesis? *Social Science Quarterly, 75*(2), 245–269.

Brulle, R. J. (1996). Environmental discourse and social movement organizations: A historical and rhetorical perspective on the development of U.S. environmental organizations. *Sociological Inquiry, 66,* 58–83.

Dunlap, R. E. (1992). Trends in public opinion toward environmental issues: 1965–1990. In R. E. Dunlap & A. G. Mertig (Eds.), *American environmentalism* (pp. 89–116). Washington, D.C.: Taylor & Francis.

Dunlap, R. E. (2000). American have positive image of the environmental movement. *Gallup Poll Monthly,* Apr., No. 415, 19–25. (Also retrieved Oct. 2000 from http://www.gallup.com/poll/releases/pr000418.asp).

Flacks, R. (1988). *Making history: The radical tradition in American life.* New York: Columbia University Press.

Holland, D. C., Lachiotte, W., Skinner, D., & Cain, C. (1998). *Identity and agency in cultural worlds.* Cambridge, Mass.: Harvard University Press.

Kempton, W., Boster, J. S., & Hartley, J. A. (1996). *Environmental values in American culture.* Cambridge, Mass.: MIT Press.

Kempton, W., Holland, D. C., Bunting-Howarth, K., Hannan, E., & Payne, C. (2001). Local environmental groups: A systematic enumeration in two geographical areas. *Rural Sociology, 66*(4), 557–578.

Kitchell, A., Hannan, E., & Kempton, W. (2000). Identity through stories: Story structure and function of two environmental groups. *Human Organization, 59*(1), 96–105.

Kitchell, A., Kempton, W., Holland, D., & Tesch, D. (2000). Identities and actions within environmental groups. *Human Ecology Review, 7*(2), 1–20.

Kuhn, M., & MacPartland, T. S. (1954). An empirical investigation of self-attitudes. *American Sociological Review, 19*(1), 68–76.

Melucci, A. (1988). Getting involved: Identity and mobilization in social movements. *International Social Movement Research, 1,* 329–348.

Stern, P. C., Dietz, T., Abel, T., Guagnano, G. A., & Kalof, L. (1999). A value-belief-norm theory of support for social movements: The case of environmentalism. *Human Ecology Review, 6*(2), 81–97.

About the Contributors

Maureen Austin is a research fellow in the School of Natural Resources and Environment at the University of Michigan in Ann Arbor. Her current research focuses on residents' perception, use, and management of nearby nature in communities with open space. Her research in urban and community forestry includes citizen participation, the exchange of information between professionals and the public, and program evaluation.

Elfriede Billmann-Machecha is a professor of psychology in the Faculty of Education at the University of Hanover in Germany, where she has been working on developing the use of group discussions as a qualitative research method for investigating how children think in different subject areas.

Amara Brook is a doctoral candidate at the University of Michigan in Ann Arbor. Her research focuses on social identity and persuasion, and how these can be used to promote individual and institutional behavior that is beneficial for the environment. She has published on wildlife management, conflict resolution, and water policy.

Susan Clayton is associate professor of social psychology at the College of Wooster in Wooster, Ohio. She has published on issues related to justice and the environment in several books and journals. With Susan Opotow, she coedited a special issue of the *Journal of Social Issues*, "Green Justice: Conceptions of Fairness and the Natural World." An earlier book (with Faye Crosby), *Justice, Gender, and Affirmative Action* (1992), received an Outstanding Book award from the Gustavus Myers Center for the Study of Human Rights.

Immo Frische is an assistant professor in the Department of Social Psychology at the University of Jena in Germany. His research interests lie in social and environmental psychology, particularly in the fields of social motivation and conflict.

Ulrich Gebhard is a professor of biology education in the Faculty of Education at the University of Hamburg in Germany. He is author of the book *Kind und Natur* (Children and Nature), a psychoanalyst for children and adolescents, and an instructor at the Institute for Psychoanalytic Art Therapy in Hanover. He has specialized in investigating children's relationships to nature and is particularly interested in the role of anthropomorphism and metaphor.

Gerhard Hartmuth is a research assistant in the Department of Economics, Sociology and Law of the UFZ-Centre for Environmental Research Leipzig-Halle in

Germany. His research interests include perception of global environmental change and the measurement of sustainability.

Dorothy Holland is professor and chair of anthropology at the University of North Carolina in Chapel Hill. Over the past decade her research has addressed personal and social agency in the context of activism, especially the environmental movement, in Nepal and the American South. Two recent books present a social practice theory of identity that she and her colleagues have been developing: *History in Person: Enduring Struggles, Contentious Practice, Intimate Identities* (edited with J. Lave, 2001) and *Identity and Agency in Cultural Worlds* (with W. Lachistte, D. Skinner, and C. Cain, 1998).

Heidi Ittner has a degree in psychology from the University of Trier, Germany and is currently at the Deutsche Forschungsgemeinschaft (German Research Foundation) in a doctoral research program on environmental psychology with a focus on mobility behavior and life-style.

Steven J. Holmes is an independent scholar of the environmental humanities, specializing in environmental life-writing. He teaches U.S. environmental history and culture at the Harvard Extension School and environmental autobiography at the Cambridge (Massachusetts) Center for Adult Education. His book *The Young John Muir: An Environmental Biography* (1999) won the Prize for Independent Scholars from the Modern Language Association, and he was awarded a 2003 fellowship from the National Endowment for the Humanities to continue his study of John Muir.

Peter H. Kahn, Jr. is research associate professor of psychology at the University of Washington in Seattle and co-director of the Mina Institute, an organization that seeks to promote, from an ethical perspective, the human relationship with nature and technology. His publications have appeared in a variety of journals. He is the author of *The Human Relationship with Nature: Development and Culture* (1999), and the coeditor, with Stephen P. Kellert, of *Children and Nature: Psychological, Sociocultural, and Evolutionary Investigations* (2002).

Linda Kalof is professor of sociology at Michigan State University. She has written extensively on issues of gender, sexuality, race, environmental values, and the links between nature and culture. She is currently the editor of *Human Ecology Review*, which publishes research on the interaction between humans and the environment.

Elisabeth Kals is *Hochschuldozentin* (associate professor) at the University of Trier in Germany. Her main current research interests cover educational psychology with a focus on communication and the applied disciplines of environmental and health psychology.

Rachel Kaplan is Samuel T. Dana Professor of Environment and Behavior in the School of Natural Resources and Environment at the University of Michigan in Ann Arbor, where she is also professor of psychology. Her work has focused on the role of nearby natural settings in people's effectiveness, satisfactions, and psychological well-being. She has also studied human needs (such as competence, identity, and being needed) in the context of environmental stewardship. Together with Stephen Kaplan she has co-authored four books, most recently *With People in Mind: Design and Management of Everyday Nature* (1998).

Willett Kempton is a cognitive anthropologist who is an associate professor in the College of Marine Studies at the University of Delaware in Newark, and is senior policy scientist in the university's Center for Energy and Environmental Policy. Kempton's research spans topics as diverse as the cultural models and values that citizens and policy makers apply to environmental issues and factors that move citizens to environmental action. He has written one book on theoretical cognitive anthropology, edited three volumes on energy conservation, and co-authored *Environmental Values in American Culture* (1995), a study of Americans' environmental beliefs and values.

Volker Linneweber holds a chair in social psychology at Magdeburg University in Germany, and teaches social psychology, personality, and environmental psychology. His research areas include interpersonal processes, with a focus on bargaining, negotiation, and conflicts. In environmental psychology, he studies evaluations of built environments and resource use as well as analyses of user needs and postoccupancy evaluations. At present, he heads the environmental psychology division of the German Psychological Association.

Gene Myers is an assistant professor in the Huxley College of Environmental Studies at Western Washington University in Bellingham. He has studied children's relationships with animals for over 14 years and authored the first scholarly and empirical book on the topic: *Children and Animals: Social Development and Our Connections with Other Species* (1998). His research examines relations with nature from a developmental perspective. Topics include the development of environmental career identity among college students, human–animal interaction, and the development of environmental caring and responsibility.

Patricia Nevers is a professor of biology education at the University of Hamburg in Germany. She has worked in genetic research for many years. She developed teaching materials for dealing with gene technology and bioethics in the classroom and worked on an empirical project that examined the role of technology in biology. Her current interests involve environmental ethics, children's views of nature, and cognitive frameworks for representing nature.

Susan V. Opotow is an associate professor in the Graduate Program in Dispute Resolution at the University of Massachusetts Boston. Her research focuses on moral exclusion, conflict, and justice in the context of environmental conflict. She has published her research and theoretical work in many book chapters and in several journals. With Susan Clayton she co-edited "Green Justice: Conceptions of Fairness and the Natural World" (1992), a special issue of the *Journal of Social Issues*. She also edited "Moral Exclusion and Injustice" (1990), a special issue of the *Journal of Social Issues*, and "Affirmative Action and Social Justice" (1993), a special issue of *Social Justice Research*.

Tarla Rai Peterson is an associate professor in the Department of Speech Communication and a research fellow in the Institute for Science, Technology, and Public Policy at the George Bush School of Government and Public Service, Texas A&M University in College Station. Her research and teaching center on the relationship between discourse and environmental policy, particularly in the context of natural resource management. She has published in the areas of environmental communication, conflict management, and rhetorical theory and criticism. She has designed and conducted training in conflict management for agencies that manage natural resources, collaborative learning workshops for citizen groups, and international workshops to develop a research agenda for public involvement in conflicts over genetically altered foods.

Linda L. Putnam is director of the Program on Conflict and Dispute Resolution in the Institute for Science, Technology, and Public Policy at the George Bush School of Government and Public Service, Texas A&M University in College Station. She is also professor of organizational communication in the Department of Speech Communication at Texas A&M University and co-editor of four books, including *Communication and Negotiation* (1992). She has published more than eighty-five articles and book chapters on conflict management, negotiation, mediation, and organizational communication. As part of the Environmental Framing Consortium, she has made presentations to dispute resolution specialists from the U.S. Environmental Protection Agency, the U.S. Department of the Interior, and the Department of the Navy.

Ann Russell is a graduate student in geography at Western Washington University in Bellingham. She has worked extensively on black bear field research projects in the Adirondack and Cascade Mountains, and has experience managing problematic interactions between bears and people. Her research involves people's perceptions of bears as subjective beings.

Charles D. Samuelson is an associate professor in the Department of Psychology and a research fellow in the Institute for Science, Technology, and Public Policy at the George Bush School of Government and Public Service, Texas A&M University in College Station. His current research interests include behavioral decision making in common-pool resource dilemmas, management of public participation in environmental policy making, and conflict management with computer-mediated communication in small groups. His research has appeared in a number of journals.

Robert Sommer is professor of psychology and chair of the art department at the University of California, Davis. A pioneer in environmental psychology, his books include *Personal Space: The Behavioral Basis of Design, Design Awareness, Tight Spaces: Hard Architecture and How to Humanize It, The Mind's Eye, Street Art, Social Design*, and *A Practical Guide to Behavioral Research* (with Barbara Sommer). He received a career research award from the Environmental Design Research Association, a research award from the California Alliance for the Mentally Ill, the Kurt Lewin Award from the Society for the Psychological Study of Social Issues, a Fulbright fellowship, and an honorary doctorate from Tallinn Pedagogical University in Estonia. For 10 years he was engaged in a cooperative project with the U.S. Forest Service that examined residents' attitudes toward city trees.

Stephen Zavestoski is an assistant professor of sociology at the University of San Francisco in California. His current research is examining the role of science in disputes over the environmental causes of unexplained illnesses, and the use of the Internet as a tool for enhancing public participation in federal environmental rulemaking. His work appears in a variety of journals and in the book *Sustainable Consumption: Conceptual Issues and Policy Problems* (2001).

Index